DOGE

(Department of Government Efficiency)

THE REBIRTH OF A NATION

AIMQWEST Corporation

AIMQWEST BOOKS
AIMQWESTBOOKS.COM

First Edition

ISBN: 979-8-218-24822-2

Prelude: A Nation at the Crossroads

In the ebb and flow of history, moments of profound transformation are often catalyzed by crises that challenge the very foundations of governance. Today, the United States finds itself at such a juncture—a nation weighed down by inefficiencies, paralyzed by bureaucracy, and burdened by a staggering national debt. Yet within this moment of peril lies the spark of possibility: the opportunity to reimagine governance, to renew trust, and to redefine what it means to serve the people.

This book, *DOGE: The Rebirth of a Nation*, is not merely a chronicle of reform but a manifesto for a new era of government efficiency and accountability. It explores the bold creation of the Department of Government Efficiency (DOGE) as a transformative force, born out of necessity and fueled by the innovative spirit of leaders like Elon Musk and Vivek Ramaswamy. It delves into the challenges of dismantling entrenched inefficiencies and replacing them with a system that is agile, transparent, and equitable.

More than a historical account, this work is a call to action. It urges policymakers, citizens, and innovators alike to engage in the vital task of reshaping governance for the 21st century. It invites readers to consider the delicate balance between efficiency and integrity, the promise of technology as a tool for equity, and the essential role of public participation in sustaining democratic ideals.

As you turn the pages of this book, you will journey through the complexities of reform and the triumphs of innovation. You will encounter the human stories behind the statistics, the ethical dilemmas behind the policies, and the visionaries who dared to imagine a better future. Above all, you will witness the rebirth of a nation poised to reclaim its promise, proving that with courage and ingenuity, even the most entrenched systems can be transformed.

Let this prelude serve as an invitation—to dream boldly, to act decisively, and to join in the effort to forge a government that is truly of the people, by the people, and for the people. The road ahead is not without challenges, but the rewards of renewal are boundless. Together, we stand on the cusp of a new chapter in our nation's story, one where governance rises to meet the needs and aspirations of all.

TABLE OF CONTENTS

Preface: The Call for Change

The Inefficiency Crisis

The inefficiency crisis gripping the nation is not merely a contemporary dilemma but a culmination of decades of incremental stagnation within the federal apparatus. It manifests in bloated bureaucracies, redundant processes, and an inability to adapt swiftly to the dynamic demands of modern governance. For citizens, this translates into delayed services, opaque decision-making, and a rising national debt that threatens economic stability. It is a crisis that imperils both public trust and the structural integrity of government institutions.

The roots of this crisis can be traced back to the unchecked growth of federal agencies, each layer of bureaucracy added ostensibly to address specific challenges, yet cumulatively creating a labyrinth of inefficiency. Overlapping responsibilities among agencies foster confusion rather than clarity, while antiquated protocols impede innovation. Reports of wasteful spending, such as unused infrastructure or outdated technology contracts, surface regularly, yet the mechanisms to address these issues remain largely ineffective. The system seems designed to perpetuate its inertia rather than facilitate meaningful reform.

Compounding these inefficiencies is the relentless ascent of national debt, now at historic levels. Decades of fiscal policy have been marked by reactive rather than proactive measures, leading to a compounding of financial obligations with no clear roadmap for resolution. This growing debt burden not only constrains economic possibilities but also undermines global confidence in the nation's fiscal discipline. Each dollar spent inefficiently within the system represents not just a loss but an erosion of potential—funds that could have been deployed toward innovation, education, or public welfare.

The urgency for reform has never been more apparent. The digital age demands a government that is agile, transparent, and accountable, characteristics that are antithetical to the status quo. The question is no longer whether reform is necessary but rather how to implement it in a way that preserves the core values of governance while embracing the efficiencies of modern technology and streamlined operations.

Against this backdrop, the establishment of the Department of Government Efficiency represents a pivotal moment in the nation's governance narrative. The department embodies a bold vision to address the inefficiency crisis head-on by leveraging technology, fostering transparency, and embracing accountability. It seeks to dismantle entrenched inefficiencies and replace them with systems that prioritize effectiveness and fiscal responsibility. By integrating advanced analytics, machine learning, and data-driven decision-making, this initiative aims to transform the very architecture of federal operations.

The stakes of addressing this inefficiency crisis are monumental. It is a battle not only to restore functionality to a broken system but also to renew public faith in the institutions that underpin democracy. Every step toward efficiency is a step toward rebuilding a nation that can meet the challenges of the twenty-first century with confidence and integrity. This is the foundational premise upon which the narrative of DOGE unfolds—a call to action for a nation poised at the brink of transformation.

The Rise of Bureaucratic Stagnation

The phenomenon of bureaucratic stagnation, an entrenched state of inertia within the administrative machinery, is a crisis that has quietly gained momentum over decades. Far from a sudden or isolated problem, it is the result of cumulative inefficiencies layered across generations of governance. Each well-intentioned initiative or expansion of federal agencies has inadvertently added complexity to an already sprawling system. While every new regulation, department, or protocol might have addressed a specific challenge at its inception, the broader effect has been to create a labyrinthine structure that resists agility and innovation.

At the heart of this stagnation lies a fundamental misalignment between the structure of the government and the dynamic nature of societal needs. Federal agencies, bound by rigid hierarchies and antiquated workflows, struggle to respond effectively to crises or adapt to technological advancements. For instance, in an era where data analytics could streamline decision-making processes, many agencies continue to rely on outdated methods, bogging down critical operations with unnecessary redundancies. Communication across departments is often hindered by silos that prevent the seamless exchange of information, leading to duplicated efforts and wasted resources.

The consequences of this stagnation ripple far beyond the confines of government offices. For citizens, it manifests as delays in services, prolonged approval times for basic applications, and a pervasive sense of frustration. Entrepreneurs encounter roadblocks when navigating convoluted licensing systems, while researchers and innovators face slow-moving approval processes for grants or permits. Each layer of inefficiency translates into a tangible cost—not only in financial terms but in the erosion of public trust in government institutions.

This issue is further compounded by the culture within many federal agencies. While there are countless dedicated civil servants who work tirelessly to serve the public, they often do so within a framework that discourages innovation and rewards adherence to established norms. Risk-aversion, a natural byproduct of bureaucracy, perpetuates a cycle in which bold solutions are stifled in favor of maintaining the status quo. Such an environment not only hinders progress but also discourages talented individuals from entering public service, thereby depriving the system of the fresh perspectives and energy it so desperately needs.

Historical attempts to address these inefficiencies have largely been piecemeal, targeting symptoms rather than the root causes of stagnation. While commissions and task forces have periodically identified opportunities for reform, their recommendations often falter in the face of political resistance or institutional inertia. Without systemic change, these efforts have been unable to achieve lasting impact, leaving the fundamental challenges unresolved.

The rise of bureaucratic stagnation is a reminder of the unintended consequences of neglecting structural evolution. As society advances in complexity, the mechanisms of governance must also evolve to remain relevant and effective. Recognizing this reality is the first step toward addressing the deep-seated inefficiencies that have taken root. The vision for reform is not merely about improving operational efficiency; it is about restoring the fundamental purpose of government—to serve its people with integrity, agility, and accountability. Only by confronting the pervasive stagnation can the foundation be laid for a system capable of meeting the challenges of the modern era.

National Debt and Fiscal Alarm Bells

As of November 2024, the U.S. national debt stands at a staggering $36.08 trillion, according to the U.S. National Debt Clock (usdebtclock.org). This immense figure, equivalent to approximately $106,811 per citizen or $273,132 per taxpayer, highlights the pressing urgency of addressing the nation's fiscal crisis. The sheer scale of this debt underscores systemic issues within government finance, where expenditures have consistently outpaced revenues for decades. It is not merely an economic statistic but a reflection of the profound challenges facing governance in modern America.

The growth of the national debt is a cumulative outcome of political decisions, economic policies, and historical exigencies. Successive administrations, irrespective of party lines, have faced the delicate balancing act of funding public programs, responding to crises, and managing economic stability. However, this balance has often tilted toward borrowing as an expedient solution. While borrowing can be a necessary tool for addressing immediate needs, such as war efforts or pandemic relief, the long-term reliance on debt financing has created a structural imbalance that jeopardizes the nation's economic future.

One of the most striking aspects of this crisis is the growing burden of interest payments on the debt, which consumes an increasing share of federal resources. In fiscal year 2024, interest payments alone accounted for hundreds of billions of dollars, surpassing expenditures on several critical programs. This drain not only limits the government's capacity to invest in essential areas such as infrastructure, education, and healthcare but also poses a direct challenge to fiscal sustainability. Each dollar spent on interest is a dollar diverted from improving the lives of citizens.

Compounding this problem is the political inertia surrounding fiscal reform. Attempts to address the debt often fall victim to partisan gridlock, where competing priorities and short-term political gains overshadow the imperative for long-term solutions. Proposals for increasing revenue through tax reforms or reducing expenditures face intense resistance, leaving the issue unresolved. This reluctance to act decisively has allowed the debt to grow unchecked, amplifying its impact on economic stability and national security.

The implications of such a massive debt burden extend beyond domestic borders. The United States, as a global economic leader, must maintain the confidence of international markets and investors. A deteriorating fiscal outlook could undermine the dollar's position as the world's reserve currency, increase borrowing costs, and destabilize global financial systems. Domestically, taxpayers bear the brunt of this crisis, facing the prospect of higher taxes and reduced public services as the government grapples with the mounting costs of its obligations.

The Department of Government Efficiency (DOGE) emerges as a vital initiative in this context, tasked with reimagining fiscal responsibility through innovative approaches and systemic reforms. By leveraging advanced data analytics and artificial intelligence, DOGE aims to identify inefficiencies, eliminate wasteful spending, and optimize resource allocation. This is not merely about cutting costs but ensuring that government expenditures align with tangible outcomes that benefit the public.

Addressing the national debt requires bold leadership, bipartisan cooperation, and a willingness to embrace transformative change. It is a challenge that transcends individual administrations or political ideologies, demanding a collective commitment to securing the nation's financial future. As the numbers on the debt clock continue to rise, they serve as a stark reminder of the urgent need for action. The time has come to confront this crisis with the resolve and innovation it demands, ensuring that the United States remains a beacon of economic resilience and governance for generations to come.

Why the Time for Reform is Now

The call for reform is no longer a distant echo but a resonant imperative reverberating through every stratum of society. The challenges posed by bureaucratic inefficiency, fiscal instability, and the erosion of public trust demand immediate action. This moment in history is unique, presenting both a convergence of crises and an unparalleled opportunity for transformation. It is a crossroads where the nation must decide whether to persist with outdated systems or embrace innovation and accountability to build a future rooted in efficiency and equity.

Several factors underscore why reform cannot be deferred. First, the magnitude of inefficiency within federal institutions has reached a tipping point, where inaction poses a direct threat to governance itself. Delayed responses to crises, opaque decision-making processes, and the staggering accumulation of waste have eroded public faith. Citizens increasingly view the government as an entity incapable of meeting their needs, a perception that undermines the very foundation of democracy. This loss of confidence, once entrenched, can take generations to restore, making timely intervention critical.

Additionally, the national debt, now towering at $36 trillion, casts a long shadow over the nation's economic future. This figure is not an abstraction; it represents a profound challenge to fiscal sovereignty, economic stability, and the ability to invest in essential services. Each day that reform is postponed, the cost of addressing this debt grows exponentially, compounding interest payments and narrowing fiscal options. Without decisive action, the nation risks being consumed by the weight of its obligations, leaving future generations burdened by choices they did not make.

Technological advancement offers a glimmer of hope amid these challenges, but it also raises the stakes for reform. The digital age demands systems that are agile, responsive, and transparent—qualities antithetical to the entrenched practices of a bureaucracy that has resisted modernization. Leveraging tools such as artificial intelligence, data analytics, and automation could revolutionize governance, enabling more efficient allocation of resources and precise policymaking. However, these innovations require a framework that prioritizes adaptability and accountability, elements currently lacking in the federal structure.

Political dynamics further amplify the urgency. Public discontent is growing, fueled by frustration with inefficiency, corruption, and partisanship. This dissatisfaction creates a volatile environment where the demand for change is increasingly vocal and widespread. Failing to act risks exacerbating societal divides, as citizens perceive the government as either unwilling or incapable of addressing their concerns. Reform, therefore, is not just an administrative necessity but a political imperative, vital to restoring legitimacy and fostering unity.

Finally, history offers a sobering reminder that moments of inertia often precede decline. Civilizations and institutions that fail to adapt to changing circumstances inevitably falter. The lessons of past reform efforts, such as the Grace Commission, highlight both the potential and the pitfalls of transformative initiatives. They underscore the importance of bold, systemic change driven by visionary leadership and sustained by public engagement.

The time for reform is not tomorrow or the next electoral cycle; it is now. This is a moment that demands courage, creativity, and collaboration. By embracing the principles of efficiency, transparency, and accountability, the nation has the chance to rebuild its institutions, reaffirm its values, and reclaim its promise. The path forward is clear: reform is not merely an option but a necessity, a call to action that must be answered with unwavering resolve.

DOGE's Vision and Mandate

The Department of Government Efficiency is a bold initiative, embodying a vision that seeks not just to reform, but to redefine the principles and practices of governance in the United States. At its heart lies an unwavering commitment to aligning governmental operations with the needs of the twenty-first century. This vision is rooted in the belief that efficiency, transparency, and accountability are not merely administrative ideals, but foundational elements of a government that serves its people effectively and responsibly.

DOGE's mandate is multifaceted, addressing inefficiencies that have plagued federal agencies for decades while charting a path toward innovation and fiscal prudence. It is tasked with eliminating wasteful expenditures that have become an entrenched feature of bureaucratic operations, ensuring that every dollar spent contributes meaningfully to public welfare. This requires not only identifying areas of inefficiency but also implementing systemic changes that prevent their recurrence. From duplicative processes to outdated technologies, DOGE is committed to overhauling the very architecture of federal operations.

Central to this mission is the integration of advanced technologies, particularly artificial intelligence and data analytics. These tools are not just means to streamline processes but are envisioned as transformative forces that will redefine how decisions are made. By leveraging real-time data and predictive analytics, DOGE aims to create a dynamic system capable of adapting to changing circumstances with precision and foresight. This technological foundation is a testament to the department's forward-looking approach, ensuring that reforms are not only effective in the present but resilient in the face of future challenges.

The vision of DOGE extends beyond efficiency; it encompasses a profound reimagining of government accountability. Transparency is a cornerstone of this initiative, with mechanisms designed to make government operations accessible and understandable to the public. By fostering a culture of openness, DOGE seeks to rebuild trust in institutions that have long been perceived as opaque and unresponsive. This commitment to transparency is coupled with an emphasis on participatory governance, inviting citizens to play an active role in shaping the policies and practices that affect their lives.

Leadership is pivotal to realizing this vision, and the appointment of figures such as Elon Musk and Vivek Ramaswamy underscores the innovative spirit driving DOGE. Musk's advocacy for technological innovation and Ramaswamy's focus on public engagement bring a unique confluence of entrepreneurial energy and civic-mindedness to the department. Together, they embody the values at the heart of DOGE: a relentless pursuit of excellence, a dedication to the public good, and a willingness to challenge the status quo.

DOGE's mandate is ambitious, reflecting the scale of the challenges it seeks to address. It is not simply about cutting costs or streamlining operations; it is about creating a government that is agile, responsive, and aligned with the aspirations of its people. By embracing this vision, the Department of Government Efficiency aims to lead the United States into a new era of governance, one defined not by its limitations, but by its possibilities. Through its bold reforms and unwavering commitment to innovation, DOGE stands as a beacon of hope, demonstrating that meaningful change is not only possible but imperative.

Roots in Historical Reforms (Grace Commission)

The Department of Government Efficiency draws inspiration from historical efforts to address inefficiencies in governance, with the Grace Commission serving as a pivotal example. Convened by President Ronald Reagan in 1982, the Grace Commission was an ambitious undertaking aimed at scrutinizing the federal government's operations to identify and eliminate waste. This initiative marked one of the most comprehensive reviews of federal spending in American history, offering valuable lessons for contemporary reform efforts.

The Grace Commission was born out of a growing frustration with government inefficiency and the ballooning federal deficit of the early 1980s. Composed of private-sector leaders, business executives, and policy experts, the commission operated independently of the federal bureaucracy, enabling it to approach its task with a fresh perspective. Its mandate was to identify inefficiencies, propose actionable solutions, and recommend cost-saving measures without compromising the quality of public services.

Over two years, the Grace Commission conducted an exhaustive review of government operations, producing a final report that outlined over 2,400 recommendations. These proposals spanned a wide array of areas, from procurement processes and personnel management to the elimination of redundant programs. Collectively, the commission estimated that its recommendations could save the federal government more than $424 billion over three years. However, the report also highlighted systemic challenges, such as entrenched bureaucratic resistance and the political difficulty of implementing significant reforms.

While many of the Grace Commission's recommendations were never fully implemented, its legacy endures as a testament to the potential for private-sector approaches to drive public-sector efficiency. The initiative demonstrated the value of leveraging expertise from outside traditional government channels, fostering innovation, and challenging the status quo. It also underscored the importance of aligning reform efforts with broader societal values, ensuring that cost-cutting measures did not come at the expense of equity or essential services.

The lessons of the Grace Commission resonate strongly with the mission of DOGE. Like its predecessor, DOGE seeks to transcend the limitations of traditional government structures by adopting a results-oriented approach to governance. Its reliance on advanced technologies, such as artificial intelligence and data analytics, builds upon the analytical rigor exemplified by the Grace Commission, while pushing the boundaries of what is possible in the digital age.

Moreover, the Grace Commission highlights the importance of transparency and public engagement in any reform effort. By making its findings widely available and advocating for bipartisan support, the commission sought to build a broad coalition for change. This emphasis on openness is mirrored in DOGE's commitment to participatory governance, where citizens play an active role in shaping the direction of reform. By fostering trust and collaboration, DOGE aims to overcome the institutional inertia that has historically stymied meaningful progress.

As the Department of Government Efficiency charts its course, the Grace Commission serves as both a guide and a cautionary tale. Its successes and shortcomings provide invaluable insights into the challenges of implementing systemic change within a complex federal apparatus. By learning from this historical precedent, DOGE is poised to achieve what the Grace Commission could only envision: a government that is leaner, more transparent, and better equipped to serve the needs of its people in an increasingly interconnected and dynamic world.

The Role of Technology and Public Engagement

The interplay between technology and public engagement represents the cornerstone of a transformative vision for modern governance. The Department of Government Efficiency understands that addressing the inefficiencies and inertia of federal systems requires tools and strategies that are not only innovative but also deeply interconnected with the people they serve. This dual focus on technological innovation and participatory governance is the key to reshaping how government operates in an increasingly digital and interconnected world.

Advanced technology, particularly artificial intelligence, stands at the heart of this transformation. AI is more than a tool for automation; it is a catalyst for a new paradigm of decision-making. By harnessing the power of machine learning, data analytics, and predictive modeling, DOGE can move beyond reactive governance to a model driven by precision and foresight. These technologies enable real-time monitoring and optimization of resources, allowing agencies to address issues proactively rather than reactively. From identifying and eliminating inefficiencies to ensuring that public funds are allocated where they are most needed, AI provides the foundation for a government that is as adaptive as it is effective.

Equally important is the role of digital platforms in fostering transparency and accountability. By making government data accessible and understandable to the public, DOGE aims to dismantle the opacity that has long plagued federal operations. Citizens can track the progress of initiatives, review spending patterns, and provide feedback, creating a continuous loop of accountability that bridges the gap between government actions and public expectations. This transparency is not just a safeguard against inefficiency or corruption; it is a means of restoring trust in institutions that have often been perceived as distant and unresponsive.

However, technology alone is insufficient without the active engagement of the people it serves. Public participation is not merely an accessory to governance; it is its lifeblood. DOGE envisions a government that thrives on collaboration, where citizens are partners in the policymaking process. Town halls, digital suggestion boxes, and online forums are just the beginning. These mechanisms provide platforms for dialogue, ensuring that diverse perspectives are not only heard but integrated into decision-making. Such participatory governance transforms citizens from passive recipients of policy into active contributors to the future of their nation.

This approach also includes leveraging the power of social media and other digital tools to engage with younger, tech-savvy generations. By meeting people where they are, DOGE can foster a culture of civic involvement that transcends traditional boundaries of age, geography, and socioeconomic status. In doing so, it creates a more inclusive and dynamic public sphere, one where innovation is not confined to technology but extends to the very fabric of civic life.

The integration of technology and public engagement is not without its challenges. Safeguarding data privacy, ensuring equitable access to digital tools, and maintaining the integrity of participatory processes are critical concerns that must be addressed. Yet, these challenges are not insurmountable. With thoughtful design and a commitment to ethical principles, DOGE can navigate these complexities, ensuring that its initiatives not only succeed but set a precedent for governance in the digital age.

At its core, the synergy between advanced technology and public engagement represents a profound reimagining of what government can be. It is a vision of a government that is not only efficient but also deeply connected to its citizens—a government that listens, learns, and evolves in response to the needs and aspirations of the people it serves. Through this integration, DOGE aims to transform governance into a living, breathing entity, capable of meeting the challenges of today while laying the groundwork for a brighter, more inclusive future.

Embracing Efficiency Without Sacrificing Integrity

Efficiency, as envisioned by the Department of Government Efficiency, must be pursued with a steadfast commitment to preserving the moral and structural integrity of governance. The balance between streamlining operations and safeguarding the foundational principles of equity, accountability, and transparency is not merely a challenge; it is the defining test of meaningful reform. For efficiency to be truly transformative, it must operate within an ethical framework that protects the values and aspirations of the people it seeks to serve.

Central to this endeavor is the understanding that efficiency cannot exist in a vacuum. It must be rooted in the recognition that government is not a business driven solely by profit margins but an institution tasked with upholding justice, ensuring fairness, and fostering public trust. The drive to eliminate waste and optimize processes must never compromise the rights and dignity of individuals. Each reform must be scrutinized not only for its operational benefits but also for its impact on the social fabric of the nation.

This principle becomes particularly salient in addressing workforce reductions and restructuring. While efficiency initiatives often necessitate downsizing or reallocating resources, these actions must be guided by compassion and foresight. Supporting displaced workers through retraining programs, career counseling, and transitional assistance ensures that reforms do not come at the expense of human well-being. The aim is not merely to cut costs but to create a system where every individual has the opportunity to contribute meaningfully to society.

Equally important is the need for transparency in implementing efficiency measures. Without openness, even the most well-intentioned reforms risk being perceived as opaque maneuvers that erode public trust. By clearly communicating the goals, processes, and outcomes of efficiency initiatives, DOGE can foster an environment of collaboration and understanding. Transparency not only holds the government accountable but also empowers citizens to engage actively in shaping their governance.

Integrity also demands vigilance against conflicts of interest. As DOGE relies on advanced technologies and private-sector expertise to achieve its goals, it must ensure that these partnerships prioritize the public good over private gain. Establishing robust oversight mechanisms, enforcing strict ethical guidelines, and maintaining a culture of accountability are essential to preserving the integrity of the reform process.

In embracing efficiency, DOGE recognizes that the true measure of success lies not in the speed or cost savings of its initiatives but in their ability to enhance the lives of citizens. Every policy, program, and reform must be evaluated for its contribution to a government that is equitable, responsive, and just. Efficiency, when guided by integrity, becomes more than a tool for cutting costs—it becomes a means of elevating governance to its highest potential.

Through this balance, DOGE seeks to set a new standard for reform, demonstrating that efficiency and integrity are not mutually exclusive but inherently interconnected. By upholding this principle, the department aspires to create a legacy of governance that is not only efficient but also profoundly ethical, ensuring that the pursuit of progress never comes at the cost of the values that define a nation.

Why This Book Matters

In a time of profound challenges and transformative potential, the importance of this book cannot be overstated. It serves as both a critical examination of the inefficiencies that plague governance and a visionary roadmap for meaningful change. By delving into the ambitions and implications of the Department of Government Efficiency, it transcends the boundaries of traditional political discourse to explore the philosophical, ethical, and operational dimensions of reform.

The narrative underscores the urgent necessity for a governance model that aligns with the complexities of the modern era. As bureaucratic stagnation and unchecked national debt threaten the very foundations of democracy, this work offers a timely intervention. It is not merely a chronicle of dysfunction but a call to action, inviting policymakers, citizens, and thought leaders to engage in the vital work of reimagining governance.

At the heart of this exploration lies an unwavering belief in the transformative power of innovation. The book articulates how advanced technologies, such as artificial intelligence and data analytics, can revolutionize public administration. It illustrates the potential for these tools to optimize resource allocation, eliminate inefficiencies, and create systems that are not only responsive but anticipatory. This focus on technology is complemented by an equally robust commitment to participatory governance, emphasizing the role of public engagement in shaping a government that is transparent, accountable, and inclusive.

The significance of this book extends beyond its subject matter. It represents a fusion of historical insight and forward-thinking analysis, drawing lessons from past reform efforts like the Grace Commission while envisioning a future shaped by technological advancement and civic collaboration. It acknowledges the challenges inherent in such an ambitious undertaking, from political resistance to the ethical considerations of workforce restructuring. Yet, it approaches these challenges with a balanced perspective, offering practical solutions that honor the integrity and dignity of those affected.

This work is as much about inspiration as it is about analysis. By spotlighting the innovative leadership of figures like Elon Musk and Vivek Ramaswamy, it highlights the potential for visionary thinking to drive systemic change. Their emphasis on transparency, creativity, and public engagement sets a powerful example, demonstrating that the pursuit of efficiency need not come at the expense of democratic values or public trust.

Ultimately, this book matters because it dares to imagine a government that is not only more efficient but also more humane and equitable. It challenges readers to move beyond cynicism and complacency, to recognize the possibilities inherent in thoughtful reform. By framing efficiency as a moral and civic imperative, it redefines the conversation around governance, making it accessible and compelling to a broad audience.

In addressing the crises of the present, this book lays the foundation for a future defined not by the limitations of the past but by the potential of what governance can become. It is an invitation to think critically, act boldly, and embrace the transformative possibilities of innovation and accountability. Through its incisive analysis and compelling vision, it seeks to inspire a generation of leaders and citizens to take up the mantle of reform, ensuring that the ideals of democracy are not only preserved but elevated for the challenges and opportunities of the twenty-first century.

A Roadmap for Governance in the Modern Age

Modern governance faces a dual imperative: it must innovate while staying grounded in the principles that define democracy. This vision requires a roadmap for transformation, one that not only responds to the inefficiencies of the present but anticipates the demands of the future. This blueprint, articulated through the mission of the Department of Government Efficiency, seeks to balance the agility of technological advancements with the enduring need for transparency, accountability, and public trust.

The challenges of the current era demand a new framework for governance—one that transcends the incrementalism of past reforms. In an age where complexity is both a hallmark and a hurdle, government institutions must move beyond static processes and embrace dynamic systems capable of adapting to rapidly shifting circumstances. This transformation hinges on leveraging the best of modern technology. Artificial intelligence, predictive analytics, and real-time data integration are no longer luxuries; they are necessities for a government that must operate at the speed of its citizenry.

But innovation alone cannot define this roadmap. Efficiency must serve a higher purpose—ensuring that government remains a force for equity and justice. As DOGE envisions a future where public services are streamlined and bureaucracy is minimized, it simultaneously reaffirms its commitment to maintaining the integrity of democratic governance. Every reform initiative must ask: does this make government not just faster, but fairer? Does it uphold the values of access and inclusivity, ensuring that efficiency does not come at the expense of those who depend on public systems the most?

Critical to this roadmap is the alignment of leadership and public engagement. Figures like Elon Musk and Vivek Ramaswamy exemplify the innovative energy that DOGE seeks to harness. Their contributions highlight the importance of combining entrepreneurial vision with civic responsibility, ensuring that private-sector strategies enrich rather than erode public trust. At the same time, this transformation requires the active participation of citizens. Through mechanisms like open-data platforms, participatory budgeting tools, and digital town halls, DOGE aims to foster a collaborative model of governance where every voice has the potential to shape outcomes.

Historical precedents like the Grace Commission offer valuable insights into the pitfalls and potentials of ambitious reform. These efforts underscore the importance of setting realistic goals, securing bipartisan support, and maintaining a relentless focus on implementation. DOGE's roadmap builds on these lessons, prioritizing actionable steps and measurable outcomes to ensure that reform translates from aspiration to achievement.

This vision for governance in the modern age is not a simple recalibration of existing systems. It is a fundamental reimagining of what government can and should be—a dynamic, responsive, and accountable institution capable of meeting the needs of a diverse and evolving populace. By outlining a clear and actionable path forward, this roadmap invites not just policymakers, but all citizens, to participate in shaping the future of their nation. Through this collaborative effort, the principles of efficiency and integrity can become the foundation for a governance model that truly reflects the values and potential of the twenty-first century.

Inspiring Policymakers and Citizens Alike

Inspiration is the cornerstone of transformation. The success of any ambitious reform initiative hinges not merely on its structural blueprint but on its ability to galvanize those with the power to enact and sustain change. For the Department of Government Efficiency, this entails engaging policymakers and citizens alike, fostering a shared vision that transcends political divides and situates efficiency as a unifying principle.

For policymakers, the narrative of efficiency must resonate as a moral and civic imperative. It is not simply about balancing budgets or streamlining processes but about restoring trust in the institutions they serve. This restoration begins with transparency—ensuring that the operations of government are open, comprehensible, and accountable. By demonstrating that every dollar spent is directed toward tangible outcomes, policymakers can reaffirm their commitment to serving the public good. Moreover, the tools provided by DOGE, including advanced analytics and participatory platforms, empower legislators to make decisions grounded in real-time data and citizen input, enhancing both the effectiveness and legitimacy of their actions.

However, inspiration must extend beyond the corridors of power. Citizens, too, must feel invested in the process of reform. Historically, public engagement has often been relegated to the periphery of governance, reduced to episodic acts of voting or occasional protests. DOGE seeks to redefine this relationship, creating avenues for continuous interaction and contribution. Digital platforms, town halls, and collaborative forums are not just mechanisms for feedback; they are spaces where citizens can co-create solutions, ensuring that governance reflects the diverse needs and aspirations of its constituents.

This dual focus on policymakers and citizens is particularly crucial in an era marked by disillusionment and polarization. Public trust in institutions has been eroded by decades of inefficiency, partisanship, and perceived inaction. Rebuilding this trust requires not only substantive reforms but also a reimagining of the relationship between government and governed. Policymakers must demonstrate that they are not merely stewards of efficiency but advocates for equity, ensuring that the benefits of reform are distributed fairly and inclusively. Similarly, citizens must recognize their role not as passive recipients but as active participants in the governance process.

The transformative potential of this shared vision is exemplified by the leadership of figures like Elon Musk and Vivek Ramaswamy.

Their commitment to transparency, innovation, and public engagement underscores the principles that underpin DOGE's mission. By modeling a governance style that is both visionary and collaborative, they provide a blueprint for how policymakers and citizens can work together to achieve common goals.

Ultimately, the success of DOGE hinges on its ability to inspire collective action. It must bridge the gap between the ideals of governance and the realities of implementation, creating a framework where policymakers and citizens alike feel empowered to contribute to a government that is efficient, equitable, and responsive. In doing so, it not only addresses the crises of the present but also lays the groundwork for a future where governance is a shared endeavor, defined by its capacity to inspire and unite.

Prologue: A Nation on the Brink

A Fractured Federal System

The fractures within the federal system of governance are not merely incidental; they are emblematic of a broader, systemic dysfunction that has evolved over decades. This fragmentation manifests in myriad ways, from overlapping jurisdictions and conflicting mandates to redundant operations and inefficient resource allocation. The result is a system that struggles to respond cohesively to both routine governance needs and emergent crises, leaving citizens and policymakers alike grappling with the consequences.

At the heart of this disjointed structure is a labyrinth of federal agencies, each with its own set of priorities, protocols, and hierarchies. While these agencies were often established with noble intentions, the lack of synchronization among them has created an environment where inefficiency thrives. Agencies frequently duplicate efforts, failing to coordinate effectively, which leads to wasted resources and missed opportunities for impactful policymaking. This lack of alignment is particularly evident in areas such as disaster response, healthcare delivery, and regulatory enforcement, where inter-agency collaboration is not just beneficial but essential.

The consequences of this fragmentation extend beyond operational inefficiencies; they erode public trust in government institutions. Citizens interacting with the system often encounter confusion and frustration, navigating conflicting requirements, delays, and inconsistencies. Businesses seeking permits or compliance certifications find themselves entangled in a web of contradictory directives, stifling innovation and economic growth. These everyday challenges underscore a deeper issue: the inability of the federal system to present a unified, coherent face to the people it serves.

Compounding these structural inefficiencies is a culture within the bureaucracy that often resists change. Risk aversion, a natural byproduct of hierarchical governance, stifles creativity and innovation. Employees within federal agencies frequently operate under rigid protocols, leaving little room for adaptive thinking or cross-departmental collaboration. This cultural inertia perpetuates the very silos that hinder effective governance, creating a feedback loop where inefficiency becomes the norm rather than the exception.

The fractured nature of the federal system also has significant implications for fiscal responsibility. Redundancies and inefficiencies lead to unnecessary expenditures, exacerbating the already dire national debt. Programs with overlapping objectives siphon resources from other critical areas, creating a misallocation of taxpayer dollars. This financial strain is further compounded by the administrative costs of maintaining a bloated bureaucratic structure, diverting funds from essential services and investments.

Addressing these fractures requires more than incremental adjustments; it necessitates a fundamental rethinking of the federal system's architecture. The Department of Government Efficiency represents a bold initiative in this regard, seeking to dismantle silos, streamline operations, and foster a culture of collaboration and innovation. By leveraging advanced technologies and promoting participatory governance, DOGE aims to create a federal system that is not only more efficient but also more transparent, accountable, and responsive.

The journey toward repairing a fractured federal system is undoubtedly complex, demanding both political will and public support. Yet, the stakes are too high to accept the status quo. By addressing these systemic issues head-on, the United States has the opportunity to build a governance model that meets the challenges of the modern era, ensuring that its institutions serve as a cohesive and effective force for progress and equity.

Stories of Waste and Redundancy

The story of waste and redundancy within the federal government is a narrative that spans decades, embodying the systemic inefficiencies that have become a hallmark of modern bureaucracy. These instances of mismanagement are not isolated anomalies but rather recurring patterns that highlight the need for comprehensive reform. Each tale of squandered resources or overlapping operations reveals a deeper truth about a system struggling to meet the demands of a rapidly evolving society.

One vivid example is the notorious issue of redundant IT systems across federal agencies. Despite billions allocated annually for information technology, many departments operate isolated systems that fail to communicate effectively with one another. The result is a patchwork of databases and platforms, each requiring separate maintenance contracts, upgrades, and personnel. Efforts to integrate these systems often stall due to inter-agency conflicts or logistical hurdles, leaving critical information siloed and inaccessible when it is needed most.

The problem extends beyond technology. Federal procurement practices have long been a breeding ground for inefficiency and waste. From overpriced military equipment to underutilized infrastructure projects, the government has repeatedly found itself paying more for less. High-profile examples, such as the infamous case of the $640 Pentagon toilet seat, have become symbols of a procurement system rife with inefficiencies and misaligned incentives. These expenditures not only drain public coffers but also undermine trust in government stewardship of taxpayer funds.

Equally troubling are the stories of programs with overlapping missions. Multiple agencies are often tasked with addressing similar issues—whether it be environmental protection, public health, or workforce development—yet operate independently, leading to duplicated efforts and inconsistent outcomes. This lack of coordination not only wastes resources but also creates confusion for the citizens and organizations these programs are meant to serve.

These redundancies and inefficiencies are further compounded by a culture that resists change. Many federal employees operate within a framework of entrenched protocols and risk-averse decision-making. Attempts to innovate or streamline operations are often met with skepticism or outright resistance, stalling progress and perpetuating the status quo. This inertia, while not born of malice, reflects a system more invested in self-preservation than in adaptability or excellence.

The consequences of such waste are far-reaching. Every dollar misallocated is a dollar that could have been invested in education, healthcare, or infrastructure. Beyond the financial toll, the inefficiency erodes public confidence in government, fostering cynicism and disengagement. Citizens who witness these stories of mismanagement are less likely to trust that their leaders are acting in their best interests, creating a vicious cycle of disillusionment and apathy.

Confronting these stories of waste and redundancy requires more than acknowledgment; it demands a transformative approach. The Department of Government Efficiency aims to tackle these inefficiencies head-on, leveraging advanced technologies and fostering a culture of collaboration and innovation. By identifying and addressing the root causes of waste, DOGE seeks not only to save resources but to rebuild trust in the institutions that serve the public. It is through this commitment to efficiency and integrity that the government can move beyond the failures of the past and toward a future defined by effectiveness and accountability.

The Costs of Complacency

Complacency within the federal government comes at an extraordinary cost, measured not only in dollars but also in missed opportunities, eroded public trust, and the gradual decline of institutional integrity. This passive acceptance of inefficiency and stagnation, cultivated over decades, has transformed what should be a dynamic and responsive system into a cumbersome machine. The inertia of the current state is more than a byproduct of bureaucratic complexity; it is a silent adversary that undermines progress and perpetuates waste.

The financial toll of complacency is staggering. Wasteful spending, often dismissed as an unavoidable consequence of large-scale governance, accumulates into a significant drain on public resources. Each misallocated dollar represents not just a loss but a forfeited opportunity to invest in education, infrastructure, healthcare, or innovation. Programs are left underfunded, critical services deteriorate, and the burden is inevitably shifted onto taxpayers. The national debt, already a pressing crisis, continues to balloon under the weight of these inefficiencies, compromising the nation's fiscal stability and future economic growth.

Beyond the monetary implications, complacency exacts a profound social cost. Citizens increasingly perceive the government as an entity detached from their needs, a perception that fosters disillusionment and disengagement. Each instance of inefficiency, whether it be a delayed response to a natural disaster or a convoluted application process for basic services, deepens the divide between the government and the people it serves. This erosion of trust is not easily repaired and poses a direct threat to the foundations of democracy.

The cultural dimensions of complacency are equally damaging. Within federal agencies, a status quo mentality often prevails, stifling innovation and discouraging bold solutions. Employees, bound by rigid protocols and risk-averse leadership, are disincentivized from challenging inefficient practices or proposing transformative ideas. This lack of accountability creates a self-perpetuating cycle where inefficiency becomes normalized, further entrenching the system's flaws.

Complacency also undermines the government's ability to respond to crises. In moments of national emergency, the agility and effectiveness of governance are paramount. Yet, a system hampered by redundant processes and outdated technologies struggles to deliver timely and coordinated action. This sluggishness not only exacerbates the immediate crisis but also diminishes the public's confidence in the government's capacity to protect and serve.

The cumulative impact of complacency is a government that fails to meet the needs of its people and falls short of its potential. The urgency for reform cannot be overstated. Addressing these systemic issues requires a proactive and sustained effort to dismantle the structures and attitudes that perpetuate inefficiency. The Department of Government Efficiency emerges as a beacon of hope in this context, with its commitment to leveraging advanced technologies, fostering innovation, and instilling a culture of accountability and excellence.

To confront the costs of complacency is to reclaim the promise of governance—a promise that prioritizes effectiveness, equity, and the well-being of every citizen. It is an endeavor that demands courage, creativity, and an unwavering dedication to progress. Only by acknowledging and addressing these deep-seated challenges can the government move from inertia to action, paving the way for a future that is both efficient and equitable.

The Trump Administration's Gamble

The Trump administration's decision to establish the Department of Government Efficiency represents one of the boldest gambles in modern American governance. Faced with mounting national debt, widespread bureaucratic inefficiencies, and growing public discontent, this initiative signals a profound shift in how government functions are conceptualized and executed. It is an undertaking that seeks not only to address systemic flaws but also to redefine the role of governance in the twenty-first century.

At its core, this gamble reflects the administration's acknowledgment of a fundamental truth: the existing structure of federal operations is unsustainable. The sprawling bureaucracy, marked by redundant agencies and outdated processes, has long struggled to meet the demands of a dynamic and interconnected world. Reform, however, is fraught with challenges. It requires dismantling entrenched systems, navigating political resistance, and confronting the inertia of decades-old practices. By creating DOGE, the administration has chosen to confront these challenges head-on, rather than continue to defer the consequences of inaction.

The decision to place Elon Musk and Vivek Ramaswamy at the helm of this transformative effort underscores the administration's commitment to innovation and unorthodox thinking. Musk, renowned for his technological vision and disruptive strategies, brings a fresh perspective to the complexities of federal governance. His emphasis on transparency and the application of advanced technologies, such as artificial intelligence and machine learning, reflects a belief in the transformative potential of data-driven decision-making. Ramaswamy, with his entrepreneurial background and focus on civic engagement, complements this vision by emphasizing the importance of public trust and participatory governance.

Yet, this initiative is not without its risks. The scale of reform envisioned by DOGE requires not only innovative leadership but also broad-based support from Congress, federal agencies, and the public. Implementing such sweeping changes within a deeply polarized political environment poses significant challenges. Critics argue that the focus on efficiency could inadvertently undermine essential services or exacerbate inequalities, highlighting the need for a careful balance between cost-cutting measures and the preservation of equity.

The administration's gamble also rests on the assumption that advanced technologies can bridge the gap between ambition and reality. While tools like AI and predictive analytics offer unprecedented opportunities for streamlining operations, their integration into government systems is not without hurdles. Concerns over data privacy, algorithmic bias, and the ethical implications of automation must be addressed to ensure that technological innovations align with democratic principles.

Despite these challenges, the potential rewards of this initiative are transformative. If successful, DOGE could serve as a model for governments worldwide, demonstrating how efficiency, transparency, and accountability can coexist within complex administrative systems. It could restore public confidence in government, foster economic growth by reducing wasteful spending, and create a more agile and responsive federal apparatus.

Ultimately, the Trump administration's gamble is about more than reforming bureaucracy; it is about redefining the relationship between government and its citizens. By embracing innovation and committing to transparency, DOGE represents a vision of governance that is not only efficient but also deeply ethical and inclusive. It is a high-stakes bet on the capacity of government to adapt and evolve—a gamble that, if realized, could mark the beginning of a new era in American democracy.

Establishing DOGE

The establishment of the Department of Government Efficiency marked a pivotal moment in American governance, driven by an urgent need to address the systemic inefficiencies that had plagued the federal apparatus for decades. The inception of DOGE was not merely an administrative restructuring; it was a declaration of intent to transform the way government serves its people, guided by principles of innovation, accountability, and fiscal prudence. At its core, this initiative represented a bold commitment to dismantling outdated systems and creating a framework capable of meeting the demands of a rapidly evolving society.

The formation of DOGE was rooted in an acknowledgment of the failures and inefficiencies inherent in the existing system. Federal agencies, burdened by overlapping mandates and entrenched bureaucratic inertia, had long struggled to operate with transparency or agility. The need for reform was undeniable, yet previous efforts had often been stymied by political gridlock, institutional resistance, and a lack of cohesive vision. DOGE was conceived as a solution to these challenges—a centralized entity empowered to lead the charge in redefining the operational and ethical standards of governance.

From its inception, DOGE was designed to function as a bridge between traditional government practices and the innovative potential of the private sector. This philosophy was evident in its leadership, with figures like Elon Musk and Vivek Ramaswamy embodying the department's commitment to entrepreneurial ingenuity and public accountability. Their appointment signaled a break from conventional governance models, emphasizing the need for fresh perspectives and unconventional approaches to reform.

Central to the establishment of DOGE was the integration of cutting-edge technologies. The department envisioned a government powered by artificial intelligence and data analytics, capable of streamlining operations, reducing redundancies, and optimizing resource allocation. This reliance on technology was not an end in itself but a means to create a more responsive and efficient system, one that could adapt to the complexities of modern governance without sacrificing the principles of equity and justice.

Equally significant was DOGE's emphasis on participatory governance. From its earliest days, the department sought to involve citizens in the reform process, recognizing that public engagement was essential to restoring trust in government institutions. By fostering transparency and creating platforms for civic participation, DOGE aimed to bridge the gap between government and the people it serves, ensuring that reform efforts were not only effective but also inclusive and representative.

The establishment of DOGE was a bold gamble, fraught with challenges and resistance. Critics questioned its feasibility, while skeptics doubted the sustainability of its ambitious goals. Yet, the vision behind DOGE was clear: to create a government that is not only more efficient but also more ethical, transparent, and accountable. This vision continues to serve as a guiding light for the department, driving its efforts to reimagine governance for the twenty-first century and beyond.

Choosing Musk and Ramaswamy

The selection of Elon Musk and Vivek Ramaswamy to lead the Department of Government Efficiency was a deliberate and symbolic choice, reflecting the Trump administration's commitment to redefining the principles of governance through bold and unconventional leadership. Their appointments were more than personnel decisions; they were declarations of intent, signaling a willingness to disrupt the status quo and embrace innovation as the cornerstone of reform.

Elon Musk's involvement brought with it the weight of a proven disruptor. Known for his transformative work across industries, from revolutionizing space travel with SpaceX to reshaping the automotive landscape with Tesla, Musk symbolizes the fusion of vision and execution. His ethos of transparency, his insistence on accountability, and his ability to leverage technology for scalable impact made him a natural fit for the task of overhauling federal inefficiencies. Musk's focus on using data-driven decision-making tools, particularly artificial intelligence, aligns seamlessly with

DOGE's mission to streamline operations and maximize resource efficiency. By drawing from his private-sector experience, Musk represents the application of entrepreneurial strategies to the public sphere, challenging traditional bureaucratic norms.

Vivek Ramaswamy's inclusion offered a complementary dimension to the department's leadership. A respected entrepreneur and advocate for public engagement, Ramaswamy brought an acute understanding of the relationship between governance and the citizenry. His commitment to participatory democracy underscored the importance of involving the public in the process of reform. By advocating for transparency and fostering dialogue, Ramaswamy emphasized the need for government to rebuild trust with its constituents. His expertise in navigating the intersection of policy, business, and public interest ensured that DOGE's initiatives would be grounded in both ethical considerations and practical realities.

The synergy between Musk and Ramaswamy exemplified a partnership of vision and pragmatism. Together, they embodied the values that the department sought to instill across the federal government: innovation, accountability, and inclusivity. Their leadership was not without challenges. Critics questioned the implications of relying on high-profile entrepreneurs to spearhead public-sector reform, raising concerns about potential conflicts of interest and the scalability of private-sector models in government. However, the administration's gamble on these leaders was predicated on their shared ability to think beyond conventional frameworks, a quality deemed essential for navigating the complexities of systemic reform.

The decision to entrust DOGE to Musk and Ramaswamy was as much about optics as it was about strategy. Their names carried with them the promise of change, signaling to both the public and the political establishment that the administration was serious about confronting inefficiency and fostering accountability. Their appointments reinvigorated public interest in government reform, sparking conversations about the role of leadership in reshaping institutions and the balance between innovation and integrity.

Ultimately, Musk and Ramaswamy's roles within DOGE reflect a broader narrative of transformation. Their leadership exemplifies the potential for collaboration across sectors, blending entrepreneurial ingenuity with civic responsibility to create a government that is not only more efficient but also more attuned to the needs and aspirations of its people. Through their guidance, DOGE set out to demonstrate that reform is not a distant ideal but an achievable reality when visionary leadership meets unwavering commitment to public service.

The Promise of Renewal

The promise of renewal represented by the Department of Government Efficiency transcends the logistical confines of bureaucracy; it is a testament to the enduring potential for governments to evolve and adapt. This initiative is not merely about rectifying inefficiencies or cutting unnecessary costs—it embodies a deeper aspiration to restore public faith, recalibrate priorities, and reinvigorate the very principles of democracy.

At its core, the commitment to renewal is a response to years of systemic dysfunction. Bureaucracies, by their nature, are slow to change, often encumbered by institutional inertia and resistance to innovation. Over time, these tendencies calcify into stagnation, creating a system more attuned to self-preservation than service. The establishment of DOGE signals a decisive break from this pattern, offering a roadmap for transformation grounded in transparency, accountability, and the judicious use of technology.

The journey toward renewal is inherently collaborative. This effort requires the engagement of every stakeholder, from policymakers and federal employees to the citizens who rely on government services. For policymakers, it means embracing a mindset of bold reform, prioritizing systemic over superficial changes. For civil servants, it entails a cultural shift toward adaptability and innovation, where risk-taking is encouraged, and outdated practices are abandoned. For citizens, it necessitates active participation, both in shaping the vision of reform and in holding leaders accountable for its outcomes.

Central to this promise is the integration of advanced technology as a tool for reform. Artificial intelligence, data analytics, and machine learning are not mere buzzwords within the vision of DOGE—they are the pillars upon which this renewal stands. These tools enable the government to operate with unprecedented efficiency, identifying waste, predicting trends, and optimizing resource allocation in ways previously unimaginable. However, the deployment of such technologies must be tempered by ethical considerations, ensuring that the pursuit of efficiency does not come at the expense of equity or individual rights.

The broader implications of renewal extend beyond operational efficiency. By addressing inefficiencies, reducing redundancies, and fostering a culture of accountability, DOGE aims to restore public trust in government institutions. Trust, once eroded, is difficult to regain, yet it is the cornerstone of effective governance. A government that is seen as responsive, ethical, and transparent is one that can mobilize its citizens toward collective goals, bridging divides and fostering unity.

Ultimately, the promise of renewal is not confined to the administrative reforms championed by DOGE; it is about reigniting the belief that government can be a force for good. It challenges the pervasive cynicism that views government as inherently inefficient and untrustworthy, offering instead a vision of what is possible when innovation meets integrity. This is the essence of DOGE's mission: to demonstrate that renewal is not only necessary but achievable, ushering in a new era where governance is as dynamic and forward-looking as the society it serves.

Lessons from Past Reforms

Historical reform efforts offer a wealth of insights into the complexities and opportunities inherent in systemic change. The lessons gleaned from these initiatives are not merely historical footnotes; they are foundational to understanding the path forward for contemporary governance. As the Department of Government Efficiency embarks on its transformative mission, it draws heavily from the successes and missteps of past reforms to inform its strategies and avoid the pitfalls that have previously hindered progress.

One of the most instructive examples is the Grace Commission of the 1980s, a bold initiative launched during President Ronald Reagan's administration. Tasked with identifying inefficiencies within federal operations, the commission brought together private-sector leaders to provide a fresh perspective on government spending. Its comprehensive review revealed extensive waste and proposed thousands of recommendations that promised billions in potential savings. However, despite the commission's thorough analysis and ambitious proposals, many of its recommendations were never fully implemented. Political resistance, bureaucratic inertia, and a lack of enforcement mechanisms ultimately undermined its impact. The Grace Commission's story highlights the critical importance of bipartisan support, robust follow-through, and the alignment of reforms with actionable frameworks.

Other reform efforts, both domestic and international, underscore the value of transparency and public engagement in fostering successful change. In the 1990s, New Zealand undertook a sweeping reform of its public sector, emphasizing performance-based management and fiscal accountability. These reforms were notable for their reliance on clear metrics and transparent reporting, which allowed citizens to assess the government's progress and hold officials accountable. The success of this initiative demonstrates how openness can serve as a powerful tool for building trust and ensuring the sustainability of reforms.

Closer to home, the U.S. has witnessed numerous state-level initiatives aimed at improving efficiency and reducing waste. Programs like Performance-Based Budgeting in Texas and Results Washington in the Pacific Northwest offer valuable case studies in leveraging data and analytics to drive decision-making. These initiatives demonstrate how targeted, data-driven approaches can yield tangible improvements in governance, even within the constraints of larger bureaucratic systems.

The key takeaway from these historical efforts is that reform requires more than identifying inefficiencies; it demands a strategic framework that integrates innovation, accountability, and public buy-in. The Department of Government Efficiency has embraced these principles, using advanced technologies and participatory governance models to ensure its initiatives are both impactful and enduring. By learning from the successes and shortcomings of past reforms, DOGE is uniquely positioned to navigate the challenges of systemic change, transforming lessons from history into a blueprint for the future of governance.

Setting the Stage for Transformation

Setting the stage for transformation requires more than envisioning change; it demands a deliberate orchestration of conditions that make reform not only possible but inevitable. The Department of Government Efficiency stands as a testament to this principle, crafting a blueprint for systemic renewal rooted in innovation, accountability, and the recalibration of priorities. At its heart lies a conviction that the time for incremental adjustments has passed, replaced by the necessity of bold and far-reaching transformation.

The foundation for this effort begins with a recognition of the entrenched inefficiencies that have defined the federal apparatus. Decades of bureaucratic growth, often driven by well-meaning but fragmented initiatives, have resulted in a system that is both sprawling and uncoordinated. Overlapping jurisdictions, redundant processes, and outdated technologies have created an environment where waste is not the exception but the rule. To dismantle this status quo, DOGE embraces a multifaceted approach that integrates advanced tools, transparent practices, and a commitment to inclusivity.

Central to this transformative vision is the adoption of cutting-edge technologies capable of streamlining operations and eliminating redundancies. Artificial intelligence and data analytics form the backbone of DOGE's strategy, offering unprecedented opportunities to analyze inefficiencies and implement targeted solutions. These tools enable not just efficiency but also precision, ensuring that reforms are both impactful and equitable. By leveraging predictive modeling and real-time data, the department aims to create a government that is not only responsive to current needs but also anticipates future challenges.

However, technology alone cannot achieve transformation. It must be paired with a cultural shift that prioritizes transparency and public engagement. DOGE recognizes that rebuilding trust in government requires more than improved functionality; it demands an open dialogue between institutions and the people they serve. Initiatives such as public dashboards, participatory forums, and citizen feedback mechanisms are integral to this effort, fostering a sense of shared responsibility and collective ownership over the reform process.

Leadership plays a pivotal role in setting the stage for transformation. The appointments of Elon Musk and Vivek Ramaswamy exemplify the fusion of visionary thinking and practical expertise required for such an ambitious undertaking. Musk's technological acumen and Ramaswamy's focus on civic engagement reflect the dual priorities of innovation and inclusivity. Their leadership provides a model for how public and private-sector principles can converge to address the complexities of modern governance.

The groundwork for transformation also involves confronting and overcoming resistance, both political and institutional. Reform inevitably disrupts entrenched power structures and challenges established norms, making it essential to build coalitions that transcend partisan divides. DOGE's strategy includes fostering bipartisan support, engaging with stakeholders at every level, and demonstrating the tangible benefits of its initiatives. By aligning its goals with broader societal values, the department seeks to create a reform movement that is both sustainable and resilient.

In setting the stage for transformation, DOGE is not merely responding to a moment of crisis; it is laying the foundation for a new era of governance. Its approach reflects a deep understanding of the challenges ahead while maintaining an unwavering focus on the possibilities of what government can achieve. Through its commitment to efficiency, transparency, and accountability, the department offers a vision of a government that is not only functional but exemplary—a model for how institutions can evolve to better serve the people and the principles they are built upon.

Chapter 1: The Genesis of DOGE

The Birth of an Idea

The birth of the idea that would become the Department of Government Efficiency began not as a fully formed vision but as a response to a mounting recognition of systemic dysfunction. The inefficiencies that plagued federal governance were not novel; they had been the subject of countless studies, commissions, and reform proposals over decades. Yet, what made this moment distinct was the convergence of public dissatisfaction, political will, and the transformative potential of technology. It was in this crucible of urgency and opportunity that the seeds of DOGE were planted.

At its inception, the idea was fueled by the realization that incremental change could no longer suffice. The challenges facing the nation were too profound, and the machinery of government too burdened by inertia, to allow for piecemeal solutions. What was required was a fundamental rethinking of how governance could and should operate. This necessitated not just addressing inefficiencies but reshaping the culture, structure, and processes that had enabled them to persist.

The inspiration for such a bold undertaking came from an unlikely confluence of sources. Lessons from historical reform efforts, such as the Grace Commission, provided a sobering reminder of the pitfalls of ambitious initiatives that lacked follow-through or public support. Simultaneously, the successes of private-sector innovation, epitomized by leaders like Elon Musk, showcased the power of technology and entrepreneurial thinking to disrupt and reimagine established systems. The challenge was to adapt these principles to the public sector while maintaining the integrity and accountability essential to democratic governance.

The concept of DOGE was also shaped by a growing recognition of the need for transparency and public engagement. Decades of opaque decision-making and unresponsive bureaucracy had eroded public trust, creating a chasm between citizens and their government. Addressing this disconnect required not just operational efficiency but a renewed commitment to openness and inclusivity. The idea was to create a system where governance was not only more effective but also more participatory, inviting citizens to play an active role in shaping policies and priorities.

From the beginning, the architects of DOGE understood that this would be no ordinary reform effort. It would require a coalition of visionary leaders, dedicated public servants, and an engaged citizenry, united by a shared commitment to transformation. The decision to involve figures like Musk and Vivek Ramaswamy was emblematic of this approach, bringing together technological innovation and civic-minded leadership to tackle the complexities of federal governance.

The birth of DOGE was not just the creation of a new department; it was the articulation of a new philosophy of governance. It represented a belief that the challenges of the modern era demand not just better systems but better principles—principles rooted in efficiency, transparency, and accountability. This idea, ambitious yet grounded, set the stage for what would become one of the most transformative initiatives in the history of American governance.

Catalysts for Creation

The catalysts for creating the Department of Government Efficiency were deeply rooted in a convergence of societal, political, and technological factors that made the necessity of reform undeniable. The pressures that shaped this initiative were not abstract; they were palpable realities confronting a nation burdened by inefficiencies and rising public dissatisfaction. Understanding these drivers provides a lens through which the origins of DOGE can be viewed as both a response to exigent crises and an embodiment of visionary governance.

A primary force behind the creation of DOGE was the sheer scale of inefficiency embedded within federal operations. Decades of unchecked bureaucratic expansion had created a labyrinth of processes that often worked against their intended purpose. Agencies, instead of cooperating, operated in silos, fostering duplication and miscommunication. The costs of this dysfunction were staggering, not only in terms of wasted taxpayer dollars but also in the erosion of public trust. Each inefficiency became a symbol of a government struggling to fulfill its responsibilities in a rapidly changing world.

The mounting national debt added an even more urgent dimension to the call for reform. By November 2024, the debt had surged past $36 trillion, a figure that starkly illustrated the unsustainable trajectory of fiscal policy. Interest payments alone were consuming a significant portion of federal resources, leaving less room for critical investments in infrastructure, education, and healthcare. The situation demanded an entity capable of tackling waste and reining in unnecessary expenditures, creating a strong economic rationale for DOGE's establishment.

Public discontent served as another powerful catalyst. Years of encountering inefficiencies in government services had left many citizens frustrated and disillusioned. Entrepreneurs faced endless red tape, families experienced delays in accessing essential services, and communities saw valuable resources squandered on ineffective programs. This widespread dissatisfaction created a political climate ripe for bold action. Leaders recognized that addressing inefficiency was not just an administrative challenge but a moral imperative, necessary to restore faith in the institutions of democracy.

Technological advancements also played a pivotal role in shaping the vision for DOGE. The rise of artificial intelligence, data analytics, and machine learning offered unprecedented tools for identifying inefficiencies and implementing solutions. Unlike past reform efforts, which relied heavily on manual audits and subjective assessments, technology promised a level of precision and scalability previously unimaginable. These innovations made it possible to envision a government that was not only leaner but also smarter, capable of making data-driven decisions that aligned with public needs.

The leadership selected to helm this transformation was another critical factor in its inception. The decision to appoint figures like Elon Musk and Vivek Ramaswamy was both practical and symbolic. Musk's track record of leveraging technology to disrupt traditional industries and Ramaswamy's focus on public engagement embodied the dual priorities of innovation and inclusivity. Their leadership underscored a commitment to bridging the divide between private-sector ingenuity and public-sector accountability.

Together, these catalysts converged to create the perfect storm for reform. They highlighted the urgent need for an entity like DOGE, one capable of addressing the inefficiencies, rebuilding public trust, and setting the stage for a new era of governance. The creation of DOGE was not an isolated event but the culmination of decades of frustration, hope, and innovation—a testament to the enduring potential for transformation in the face of systemic challenges.

Aligning Goals with the Constitution

The alignment of DOGE's goals with the United States Constitution is a cornerstone of its mission, reflecting a deliberate effort to balance the pursuit of efficiency with the preservation of democratic principles. This alignment is neither incidental nor symbolic; it is a critical framework that ensures the department's initiatives are rooted in the foundational values of governance while addressing the pressing demands of the modern era. The Constitution, as the supreme law of the land, provides both the constraints and the opportunities within which DOGE operates, serving as a guide to ensure reforms enhance rather than undermine the democratic ethos.

At the heart of this alignment is the principle of accountability, a value deeply enshrined in the Constitution through mechanisms such as checks and balances and the separation of powers. DOGE's emphasis on transparency echoes these principles, creating systems where government actions are subject to scrutiny and public oversight. By making data accessible and processes clear, the department seeks to empower citizens with the tools to hold their leaders accountable, fostering a renewed trust in the institutions that govern them.

Equally central to this alignment is the commitment to equity and justice, ideals that permeate the Constitution's promises of equal protection and due process. As DOGE implements initiatives to streamline operations and eliminate inefficiencies, it does so with a conscious effort to ensure these changes do not disproportionately disadvantage any group. This requires careful analysis and ethical considerations to prevent the unintended consequences of reforms, such as exacerbating inequalities or marginalizing vulnerable populations. In this way, DOGE not only respects constitutional principles but actively works to uphold them.

Another critical aspect of this alignment is the protection of individual liberties. The Constitution's guarantees of freedoms—speech, privacy, and association—serve as a constant reminder that efficiency must never come at the cost of fundamental rights. DOGE's reliance on advanced technologies, including artificial intelligence and data analytics, is tempered by stringent safeguards to protect privacy and prevent misuse. By prioritizing ethical AI deployment and ensuring transparency in automation, the department demonstrates its commitment to harmonizing innovation with constitutional protections.

The Constitution also inspires DOGE's participatory governance model, which seeks to involve citizens directly in the decision-making processes that shape their lives. This approach is a modern extension of the constitutional ideal of popular sovereignty, where power ultimately resides with the people. Through initiatives such as public dashboards, open forums, and crowd-sourced solutions, DOGE aims to create a government that is not only more efficient but also more inclusive and reflective of the diverse voices it serves.

In aligning its goals with the Constitution, DOGE underscores its dedication to reform that is not only transformative but also principled. This alignment ensures that the department's efforts to modernize governance remain tethered to the enduring values of democracy, justice, and liberty. It is a reminder that the pursuit of efficiency, when guided by these principles, is not merely a bureaucratic exercise but a profound act of governance that reaffirms the nation's commitment to its foundational ideals. Through this alignment, DOGE aspires to achieve a balance where innovation and tradition coexist, creating a model of governance that is as forward-looking as it is faithful to its constitutional heritage.

Assembling the Leadership Team

Assembling the leadership team for the Department of Government Efficiency was a strategic and visionary effort, driven by the recognition that the success of such an ambitious initiative depended on exceptional individuals capable of navigating complex challenges. This process was not just about filling roles but about curating a team that could embody and execute the transformative goals of DOGE. The choices made in this assembly reflected a deliberate effort to combine diverse skill sets, forward-thinking perspectives, and a shared commitment to innovation and accountability.

At the heart of this team were figures like Elon Musk and Vivek Ramaswamy, whose appointments symbolized the bold, unconventional ethos of DOGE. Musk's track record as a disruptor and innovator brought an unparalleled depth of experience in leveraging technology to overcome entrenched inefficiencies. His leadership style, characterized by transparency and an insistence on data-driven decisions, aligned perfectly with DOGE's mission to modernize governance. Musk's involvement was also a statement about the department's commitment to embracing the cutting edge of technological advancement, signaling that no challenge was too complex for creative solutions.

Ramaswamy, on the other hand, represented a complementary strength in civic engagement and ethical leadership. His focus on fostering trust and ensuring inclusivity provided the human-centered counterbalance to Musk's technological rigor. Ramaswamy's expertise in navigating the intersection of business, policy, and public sentiment ensured that DOGE's initiatives would be grounded not only in innovation but also in the values of equity and transparency. Together, their leadership laid the foundation for a department that could think beyond conventional boundaries while remaining rooted in the principles of democratic governance.

The broader leadership team was assembled with a similar focus on diversity and expertise. Each member was selected not only for their technical skills but also for their ability to adapt and innovate in a rapidly changing landscape. Data scientists, policy experts, and operational strategists were brought together to create a multidisciplinary team capable of tackling the department's multifaceted challenges. This collaborative approach ensured that DOGE's strategies were informed by a wide range of perspectives, fostering a culture of creativity and resilience.

Critical to the success of this team was its ability to integrate public and private sector expertise. By bringing in leaders with experience in entrepreneurship, technology, and public administration, DOGE was able to bridge the often-daunting gap between government processes and the efficiencies of the private sector. This blend of backgrounds enabled the team to approach problems from multiple angles, ensuring that solutions were both innovative and practical.

The assembly of this leadership team also reflected a commitment to transparency and accountability. From the outset, the selection process was guided by clear criteria and open communication, reinforcing DOGE's emphasis on ethical governance. Each appointment was accompanied by a detailed explanation of the individual's qualifications and their expected contributions to the department's mission, fostering public trust in the process.

Ultimately, the leadership team of DOGE represented the embodiment of its core values: innovation, inclusivity, and integrity. This carefully curated group of individuals was not only tasked with executing the department's ambitious goals but also with setting a new standard for what leadership in government could achieve. Through their combined efforts, they laid the groundwork for a transformative era in governance, demonstrating that with the right team, even the most daunting challenges could be met with confidence and creativity.

Elon Musk: Transparency and Innovation

Elon Musk's involvement in the Department of Government Efficiency serves as both a pragmatic choice and a symbolic gesture, reflecting the department's dual commitment to transparency and innovation. His appointment was a declaration that systemic reform would not be a continuation of incremental changes but a bold leap into the future, guided by the principles of technological advancement and ethical governance. Musk, with his unparalleled track record of disrupting industries and his ethos of relentless problem-solving, brought a unique vision to the table—one that embraced complexity while seeking elegant solutions.

Transparency, for Musk, is not merely an administrative ideal but a foundational requirement for functional governance. His career has been defined by a willingness to operate in the open, whether through open-sourcing Tesla's patents or candidly addressing public concerns about SpaceX's ambitious ventures. Within DOGE, this commitment to openness translated into an insistence on demystifying governmental operations. Musk advocated for systems that allowed citizens to see, understand, and even interact with the decision-making processes of their government. Public dashboards, real-time data access, and AI-driven insights became central components of DOGE's strategy to rebuild trust in institutions that had long been perceived as opaque.

Innovation, Musk's hallmark, was equally integral to his role. He brought with him not just an understanding of cutting-edge technologies but a mindset that valued experimentation and iterative improvement. Under his influence, DOGE adopted a technological framework that prioritized efficiency and adaptability. Artificial intelligence, machine learning, and predictive analytics were deployed to identify inefficiencies and streamline operations across federal agencies. Musk's insistence on data-driven solutions ensured that reforms were not only effective but also measurable, allowing the department to continually refine its strategies based on tangible outcomes.

Musk's leadership also reflected an unorthodox approach to governance, challenging traditional hierarchies and fostering a culture of collaboration and creativity. He believed that the same principles that had driven SpaceX to land reusable rockets and Tesla to revolutionize electric vehicles could be applied to government operations. Failures, in Musk's view, were opportunities to learn and improve, a philosophy that encouraged innovation within the often risk-averse environment of federal bureaucracy.

Critically, Musk's vision extended beyond the technical aspects of governance. He understood that technology, while powerful, is only a tool. Its application must be guided by ethical considerations and a commitment to serving the public good. Musk was vocal about the need to address potential pitfalls, such as algorithmic bias and data privacy concerns, ensuring that DOGE's initiatives upheld the principles of equity and justice. This balanced approach reinforced the department's credibility, demonstrating that efficiency need not come at the expense of humanity or fairness.

Elon Musk's role in DOGE exemplifies the transformative potential of leadership that combines transparency, innovation, and ethical responsibility. His contributions laid the groundwork for a government that not only functions more effectively but also redefines its relationship with the citizens it serves. Through his influence, DOGE became a beacon of what governance could achieve when guided by vision and integrity, offering a model for the future that is as inspiring as it is attainable.

Vivek Ramaswamy: Public Engagement

Vivek Ramaswamy's leadership within the Department of Government Efficiency is marked by his unwavering commitment to public engagement, a principle he has championed throughout his career. For Ramaswamy, the success of any governmental reform hinges on the trust and participation of the people it seeks to serve. His role within DOGE reflects a deep understanding of the dynamic interplay between transparency, civic involvement, and institutional accountability.

Central to Ramaswamy's approach is the belief that governance should not only be efficient but also inclusive, ensuring that citizens are active participants rather than passive recipients. His initiatives within DOGE have focused on creating mechanisms that bridge the gap between government operations and the people they impact. Digital platforms, town halls, and collaborative forums are not merely tools for communication—they are avenues for co-creation, empowering citizens to contribute ideas and hold leaders accountable. This participatory model transforms governance from a top-down system into a dialogue, fostering a sense of shared responsibility and collective purpose.

Ramaswamy's emphasis on public engagement is rooted in his broader vision of transparency as a cornerstone of effective governance. He has been a vocal advocate for making government data accessible and actionable, believing that informed citizens are essential to a healthy democracy. Under his guidance, DOGE has implemented initiatives such as public dashboards and real-time updates on reform progress, enabling citizens to track developments and provide feedback. These tools are designed not only to enhance accountability but also to rebuild trust in institutions that have often been viewed as opaque and unresponsive.

What sets Ramaswamy apart is his ability to navigate the complexities of public sentiment and political realities. He recognizes that fostering engagement requires more than technological solutions; it demands a cultural shift within government agencies. Through training programs and leadership development initiatives, he has worked to instill a mindset of openness and responsiveness among public servants. This cultural transformation ensures that engagement efforts are not superficial but deeply integrated into the fabric of governance.

Critically, Ramaswamy's focus on engagement extends to marginalized and underrepresented communities. He has championed efforts to ensure that reform initiatives address systemic inequities and prioritize inclusivity. By actively seeking input from diverse voices, DOGE has been able to develop policies that are not only efficient but also equitable, reflecting the varied needs and aspirations of the nation's population.

Vivek Ramaswamy's leadership demonstrates that public engagement is not a peripheral concern but a central pillar of governance. His approach combines innovation with empathy, leveraging both technology and human connection to create a government that is transparent, accountable, and deeply connected to its citizens. Through his work, DOGE has redefined what it means to engage the public, proving that inclusivity and efficiency can—and must—go hand in hand in the pursuit of meaningful reform.

The Foundational Goals

The foundational goals of the Department of Government Efficiency are the bedrock of its transformative mission, reflecting a bold commitment to reshaping governance for the modern era. These goals are not simply administrative benchmarks; they are aspirational targets designed to realign the federal apparatus with its core purpose of serving the public effectively, ethically, and sustainably. Through a careful balance of innovation, accountability, and inclusivity, these objectives aim to address systemic inefficiencies while fostering a renewed trust in government institutions.

The first cornerstone of these goals is the elimination of wasteful spending, a persistent issue that has long undermined the government's ability to allocate resources efficiently. The staggering national debt, coupled with reports of unnecessary expenditures, highlights the urgent need for a comprehensive approach to fiscal responsibility. DOGE's mandate includes identifying redundancies, streamlining operations, and implementing advanced analytics to track and evaluate expenditures. By ensuring that every dollar spent translates into meaningful outcomes, the department seeks to build a leaner, more effective federal system.

Equally critical is the enhancement of transparency, a principle deeply embedded in the fabric of democratic governance. Transparency is not merely a procedural objective; it is a moral imperative that underpins public trust. For too long, the inner workings of government have been shrouded in opacity, breeding suspicion and disengagement among citizens. DOGE aims to reverse this trend by adopting tools and practices that make governance more accessible and accountable. Public dashboards, real-time reporting, and citizen-centric platforms are envisioned as mechanisms to foster openness, allowing individuals to track progress, understand decisions, and hold leaders accountable.

The creation of an agile bureaucracy represents another pivotal goal, addressing the long-standing rigidity that has stifled innovation and responsiveness within federal agencies. Bureaucratic inertia, characterized by outdated protocols and risk-averse cultures, has often prevented timely responses to emerging challenges. DOGE's focus on agility involves integrating advanced technologies such as artificial intelligence and machine learning to create a system that can adapt to changing circumstances with precision. This agility is not limited to technological solutions but extends to fostering a culture of creativity and flexibility among public servants, empowering them to innovate and take calculated risks in pursuit of excellence.

Underpinning these goals is a steadfast commitment to inclusivity and equity. DOGE recognizes that efficiency must never come at the cost of fairness or accessibility. Reforms are carefully designed to address systemic inequities and ensure that all communities, particularly those historically marginalized, benefit from the changes. This approach reflects a holistic understanding of governance, where efficiency and justice are not opposing forces but complementary pillars of effective administration.

The foundational goals of DOGE are more than a response to the inefficiencies of the past; they are a vision for the future. By eliminating waste, enhancing transparency, creating agility, and prioritizing inclusivity, the department sets a new standard for what governance can achieve. These objectives are not merely aspirations but actionable imperatives, guiding DOGE's efforts to build a government that is not only functional but transformative— a government that truly serves its people.

Eliminating Waste

Eliminating waste within the federal government is not merely a goal but a cornerstone of the Department of Government Efficiency's mission. Waste, in all its forms, represents a failure to fulfill the promise of governance—a diversion of resources that could otherwise enhance public welfare, foster innovation, and strengthen national resilience. Addressing this systemic issue requires more than identifying inefficiencies; it demands a cultural and structural transformation that redefines how resources are allocated, utilized, and accounted for.

The presence of waste within the federal apparatus has long been a symptom of deeper systemic issues. Overlapping responsibilities between agencies, outdated technological systems, and rigid bureaucratic protocols create an environment where inefficiency is not an aberration but a norm. This inefficiency manifests in tangible ways, from redundant programs that duplicate efforts to procurement processes that prioritize compliance over value. Each inefficiency compounds the other, creating a cascade of resource mismanagement that burdens taxpayers and undermines public trust.

To confront this issue, DOGE has adopted a proactive and multifaceted approach. At the core of its strategy is the application of advanced technologies such as artificial intelligence and data analytics. These tools enable the department to conduct a granular analysis of federal operations, identifying areas where resources are being misused or underutilized. Predictive analytics allow DOGE to anticipate wasteful trends before they materialize, fostering a preventative rather than reactive stance. By embracing these technologies, the department has positioned itself at the forefront of a new era in resource management.

However, technology alone is insufficient without a cultural shift toward accountability. DOGE's initiatives are designed to foster a culture of responsibility at every level of government. This includes training programs for public servants to instill a mindset of efficiency, as well as instituting clear performance metrics to evaluate the effectiveness of federal programs. By tying resource allocation to measurable outcomes, DOGE ensures that every dollar spent aligns with the overarching goal of public benefit.

The elimination of waste also requires a nuanced understanding of its root causes. Not all inefficiencies stem from negligence or mismanagement; some arise from well-intentioned policies that fail to adapt to changing circumstances. Recognizing this, DOGE places a strong emphasis on continuous improvement. Programs and processes are subject to regular evaluations, ensuring that they remain relevant and effective over time. This iterative approach prevents the entrenchment of inefficiencies, creating a dynamic system capable of evolving alongside societal needs.

Perhaps most importantly, DOGE's efforts to eliminate waste are grounded in transparency. By making data on resource allocation and program outcomes publicly accessible, the department invites citizens to participate in the process of governance. This openness not only enhances accountability but also empowers the public to act as a check on inefficiency. Through platforms that enable citizens to report instances of waste or suggest improvements, DOGE creates a participatory model of governance that strengthens the relationship between institutions and the people they serve.

The task of eliminating waste is monumental, but its rewards are equally profound. Every inefficiency addressed is an opportunity to redirect resources toward critical areas such as education, healthcare, and infrastructure. More than a fiscal responsibility, it is a moral imperative—a commitment to honoring the trust that taxpayers place in their government. Through its comprehensive and innovative approach, DOGE demonstrates that the elimination of waste is not merely an administrative challenge but a transformative act, laying the foundation for a government that is efficient, equitable, and truly accountable to its people.

Enhancing Transparency

Enhancing transparency within the federal government is one of the most transformative aspirations of the Department of Government Efficiency. Transparency is not simply a mechanism for improving accountability; it is the cornerstone of a government that seeks to rebuild trust and engage with its citizens in meaningful ways. In a time when skepticism about government motives and actions has reached unprecedented levels, this commitment to openness represents a profound shift in the ethos of public administration.

The absence of transparency has historically allowed inefficiencies to flourish unchecked. Complex bureaucratic processes, often inaccessible to the average citizen, have created a sense of disconnection between government operations and public needs. Decisions made behind closed doors, whether out of necessity or tradition, have fostered an atmosphere of suspicion. Transparency, therefore, is not just about revealing data; it is about bridging this divide and restoring faith in governance.

DOGE's approach to transparency is multifaceted, beginning with a commitment to making government data openly accessible. Through user-friendly digital platforms, citizens are now able to track federal spending, evaluate the performance of programs, and understand how decisions are made. These platforms are designed to demystify the complexities of governance, providing real-time insights into everything from budget allocations to the progress of reform initiatives. By making this information readily available, DOGE empowers the public to participate in oversight and decision-making processes in ways that were previously unimaginable.

Central to this initiative is the use of advanced technologies that enhance the accessibility and accuracy of information. Artificial intelligence and data analytics play a critical role in compiling, analyzing, and presenting data in a transparent manner. These tools enable the identification of trends, inefficiencies, and opportunities for improvement, all of which are shared with the public in a clear and actionable format. Transparency, in this context, is not a passive release of information but an active engagement with the data that shapes governance.

Beyond the digital sphere, DOGE's transparency initiatives extend to fostering direct engagement between government officials and citizens. Town halls, public forums, and participatory budgeting exercises are reimagined as spaces for dialogue and accountability. These efforts ensure that transparency is not limited to the dissemination of information but also includes creating platforms for citizens to voice their concerns, provide feedback, and influence policy decisions.

The ethical implications of transparency are equally significant. DOGE recognizes that openness must be balanced with considerations of privacy and security. Sensitive information, particularly in areas such as national security and personal data, must be handled with care to prevent misuse or unintended consequences. The department's commitment to ethical transparency ensures that the pursuit of openness does not compromise the rights or safety of individuals.

By prioritizing transparency, DOGE is not only addressing the inefficiencies of the present but also laying the groundwork for a more engaged and empowered citizenry. This approach transforms governance from a top-down system into a collaborative partnership between government and the people it serves. Transparency becomes a tool for accountability, a catalyst for trust, and a bridge to a future where governance is as open as it is effective. Through its unwavering commitment to these principles, DOGE sets a precedent for what is possible when transparency is embraced as a foundational value.

Creating an Agile Bureaucracy

Creating an agile bureaucracy is central to the transformative goals of the Department of Government Efficiency. Bureaucracies, by their very nature, are designed for stability and predictability. However, in a world characterized by rapid technological advancement and evolving societal needs, this inherent rigidity often becomes a liability. DOGE's mission to foster agility within the federal apparatus represents a paradigm shift, aiming to make governance not only efficient but also responsive and adaptable.

The necessity for agility arises from the challenges of navigating modern complexities. Traditional bureaucratic structures are often bound by hierarchical decision-making processes and outdated protocols that inhibit quick responses. This rigidity is particularly evident during crises, where slow-moving systems fail to address urgent needs effectively. To counteract this, DOGE emphasizes the integration of technology and a cultural overhaul that prioritizes flexibility and innovation.

One of the fundamental strategies for achieving an agile bureaucracy is the adoption of advanced technologies. Artificial intelligence and machine learning are at the forefront of this transformation, providing tools that can analyze vast datasets in real-time, predict trends, and streamline decision-making processes. These technologies enable a level of operational efficiency that traditional systems cannot match, allowing agencies to respond swiftly to changing circumstances. Furthermore, automation reduces the burden of repetitive tasks, freeing human resources to focus on strategic planning and innovation.

Beyond technology, agility requires a cultural shift within the government workforce. DOGE recognizes that flexibility must be embedded in the mindset of public servants. Training programs and leadership initiatives are designed to foster a culture of adaptability, encouraging employees to embrace change and take calculated risks. This approach challenges the entrenched norms of risk aversion and resistance to innovation that have long characterized bureaucratic environments.

Collaboration is another cornerstone of agility. Silos within agencies often hinder the seamless flow of information and resources, leading to inefficiencies and duplicated efforts. DOGE's initiatives aim to break down these silos, promoting interagency cooperation and cross-functional teams. By encouraging collaboration, the department ensures that knowledge and resources are leveraged effectively, enabling a more holistic approach to problem-solving.

Accountability also plays a crucial role in fostering agility. Transparent performance metrics and regular evaluations ensure that initiatives remain aligned with their objectives and are subject to continuous improvement. This iterative approach allows for real-time adjustments and ensures that reforms are not static but evolve alongside emerging challenges and opportunities.

The ultimate goal of creating an agile bureaucracy is to build a government that can anticipate and adapt to the needs of its citizens in a timely manner. Agility is not just about speed; it is about resilience, innovation, and the ability to pivot effectively in the face of uncertainty. Through its comprehensive approach, DOGE is redefining what it means to govern in the twenty-first century, proving that adaptability and efficiency are not mutually exclusive but inherently interconnected. This transformation lays the groundwork for a federal system that is not only robust but also dynamic, capable of meeting the complexities of the modern era with confidence and competence.

Overcoming Early Challenges

Overcoming early challenges was an inevitable phase in the establishment of the Department of Government Efficiency. Ambitious in scope and disruptive by design, DOGE faced significant obstacles as it sought to carve a new path for governance. These challenges were multifaceted, spanning political resistance, institutional inertia, and public skepticism. Addressing these issues was not merely about resolving logistical hurdles; it was about demonstrating the credibility and viability of a vision that promised to transform the very architecture of federal operations.

Political resistance emerged as one of the most formidable barriers. The establishment of DOGE represented a direct challenge to entrenched power structures and long-standing practices within the federal bureaucracy. Many stakeholders, wary of losing influence or autonomy, viewed the department's initiatives as a threat rather than an opportunity. This resistance was amplified by partisan divides, with critics questioning the motives and methods behind DOGE's creation. To navigate this landscape, the department employed a strategy of inclusivity and dialogue, seeking to build coalitions across ideological lines. By emphasizing the nonpartisan benefits of efficiency, such as cost savings and improved services, DOGE gradually garnered support from diverse political factions.

Institutional inertia posed another significant hurdle. Bureaucracies, by their nature, are resistant to change, often prioritizing stability over innovation. The deep-seated culture within federal agencies—characterized by risk aversion and adherence to established norms—created an environment where new ideas were met with skepticism. To counter this, DOGE implemented comprehensive training programs aimed at fostering a culture of adaptability and collaboration. These initiatives encouraged employees to embrace innovation as a core value, transforming resistance into an opportunity for growth.

Public skepticism was perhaps the most profound challenge, reflecting a broader disillusionment with government institutions. Years of inefficiency and unfulfilled promises had eroded trust, creating a perception that any reform effort was destined to fail. Overcoming this skepticism required more than rhetoric; it demanded tangible results. DOGE prioritized transparency from the outset, ensuring that every initiative was accompanied by clear metrics and open communication. By providing citizens with real-time updates on progress and outcomes, the department sought to rebuild trust through accountability and engagement.

Another critical component of addressing early challenges was managing expectations. The ambitious goals of DOGE, while inspiring, also risked creating unrealistic hopes for immediate transformation. Recognizing this, the department adopted a phased approach to reform, focusing initially on achievable milestones that demonstrated its efficacy. Early successes, such as the elimination of redundant programs and the streamlining of procurement processes, served as proof points, building momentum and credibility for broader initiatives.

Despite these challenges, the resilience and adaptability of DOGE's leadership and staff were instrumental in navigating this complex landscape. The department's ability to confront resistance with dialogue, inertia with innovation, and skepticism with results exemplified its commitment to not just achieving reform but sustaining it. These early struggles were not merely obstacles but opportunities to refine strategies and strengthen resolve, laying the foundation for a department capable of fulfilling its transformative mandate. Through persistence and ingenuity, DOGE emerged from these challenges not only intact but emboldened, ready to fulfill its promise of efficiency, accountability, and a government reimagined.

Political Resistance

Political resistance to the Department of Government Efficiency arose not as a surprise but as an expected challenge for such an ambitious and transformative initiative. The establishment of DOGE represented a direct confrontation with the entrenched norms and power structures that have long defined federal governance. As with any effort that seeks to dismantle the status quo, the department encountered opposition from individuals and institutions wary of change, concerned about perceived threats to their influence, and skeptical of the feasibility of DOGE's mission.

One of the primary sources of resistance was the fear of decentralization and the erosion of established authority. Many federal agencies, accustomed to operating autonomously within their domains, viewed DOGE's oversight as an encroachment on their jurisdiction. This sentiment was compounded by a historical tendency within bureaucracies to prioritize self-preservation, with leaders often reluctant to cede control or implement reforms that might diminish their agency's role or resources. For these stakeholders, DOGE's emphasis on eliminating redundancies and streamlining processes was seen not as an opportunity for improvement but as a potential threat to their established power dynamics.

Compounding this resistance were ideological divides that permeated the political landscape. Critics on the left and right questioned different aspects of DOGE's vision. Some conservatives argued that the department's mandate represented an overreach of federal authority, paradoxically expanding government in its quest for efficiency. Progressives, on the other hand, raised concerns about potential workforce reductions and the ethical implications of consolidating decision-making power. These ideological disagreements created a fragmented opposition, making consensus on the necessity and scope of reform elusive.

DOGE's reliance on advanced technologies such as artificial intelligence and machine learning also drew scrutiny. Concerns about data privacy, algorithmic bias, and the potential for misuse of technology fueled skepticism about the department's ability to implement reforms responsibly. Critics warned that an overreliance on automation could lead to dehumanized decision-making, undermining the very accountability DOGE sought to enhance. These fears, while not unfounded, often overshadowed the department's efforts to address them through rigorous safeguards and ethical oversight.

Navigating this resistance required a delicate balance of diplomacy, transparency, and resilience. DOGE's leadership, particularly Elon Musk and Vivek Ramaswamy, played a pivotal role in addressing these concerns. Musk's reputation as a transformative innovator and Ramaswamy's focus on public engagement served as a counterweight to the skepticism, demonstrating that the department was guided by both visionary and practical considerations. Their ability to articulate DOGE's goals in terms that resonated across political and ideological divides was instrumental in building tentative support for the initiative.

DOGE also adopted a strategy of incremental implementation, focusing initially on smaller, demonstrable successes that could serve as proof points for its broader vision. By addressing easily identifiable inefficiencies and delivering tangible results, the department sought to undermine resistance by demonstrating the practical benefits of its reforms. These early victories helped to build credibility and shift the narrative from one of doubt to one of cautious optimism.

Ultimately, the resistance to DOGE underscored the inherent difficulty of transformative change. Reform, by its very nature, challenges entrenched interests and long-standing norms, making opposition inevitable. Yet, through its commitment to transparency, ethical accountability, and dialogue, DOGE was able to navigate this resistance and lay the groundwork for a new era of governance. The department's ability to overcome these challenges not only validated its mission but also reaffirmed the resilience of democratic systems in the face of innovation and progress.

Public Skepticism

Public skepticism toward the Department of Government Efficiency emerged as a significant hurdle in its formative stages, reflecting broader societal disillusionment with promises of reform that had often failed to materialize. This skepticism was rooted in years of witnessing inefficiencies persist despite numerous commissions, initiatives, and policy adjustments purportedly aimed at improving government operations. For many, DOGE seemed like another well-intentioned effort that would succumb to the same inertia and resistance that had stymied previous attempts at transformation.

The public's doubts were not unfounded. Decades of wasteful spending, opaque decision-making, and partisan gridlock had eroded trust in the government's ability to act in the best interests of its citizens. High-profile examples of inefficiency—such as unused federal infrastructure projects or redundant administrative processes—served as potent symbols of a system seemingly incapable of self-correction. Against this backdrop, the creation of DOGE was met with cautious optimism at best, and outright skepticism at worst.

Compounding this doubt was the ambitious nature of DOGE's mandate. Promising to eliminate waste, enhance transparency, and create a leaner government, the department's goals were perceived by some as overly idealistic, if not impossible. Critics questioned whether a single entity could effectively address inefficiencies that were deeply ingrained in the federal bureaucracy. Others viewed the emphasis on technological solutions with suspicion, fearing that an overreliance on artificial intelligence and automation might exacerbate existing inequities or create new vulnerabilities.

DOGE's leadership recognized that overcoming this skepticism would require more than rhetoric or theoretical frameworks. Demonstrating tangible progress was paramount. Early initiatives focused on producing visible, measurable outcomes that could serve as proof points for the department's capabilities. By targeting low-hanging fruit, such as redundant programs or inefficient procurement practices, DOGE delivered quick wins that began to shift public perception. These successes, though modest in scale, signaled that change was not only possible but underway.

Transparency played a crucial role in addressing public doubts. From its inception, DOGE committed to making its operations and outcomes accessible to citizens. Real-time updates on reform initiatives, accessible through public dashboards, allowed individuals to track progress and understand how decisions were being made. This openness helped to counteract suspicions of secrecy or ulterior motives, fostering a sense of accountability that had been lacking in previous reform efforts.

Equally important was DOGE's emphasis on public engagement. Recognizing that skepticism often stems from a sense of disconnection, the department made concerted efforts to involve citizens in its work. Town halls, open forums, and digital platforms provided opportunities for individuals to voice their concerns, offer feedback, and contribute ideas. This participatory approach not only enhanced the department's credibility but also empowered citizens to become active stakeholders in the reform process.

Over time, these strategies began to erode the barriers of skepticism, replacing doubt with cautious confidence. The journey was not without setbacks, as challenges and criticisms persisted, but DOGE's ability to adapt and respond to public concerns demonstrated its commitment to genuine reform. By aligning its actions with its promises, the department proved that skepticism, while natural, could be overcome through transparency, engagement, and results. In doing so, DOGE laid the groundwork for a renewed trust in the institutions of governance and the belief that transformation, though difficult, is achievable.

Chapter 2: Breaking Down the Bureaucracy

Mapping the Federal Labyrinth

Mapping the federal labyrinth is an essential first step in addressing the inefficiencies that have long plagued government operations. The complexity of the U.S. federal bureaucracy is both a strength and a weakness, reflecting a system designed to manage diverse needs and responsibilities, but which has, over time, become an intricate web of overlapping mandates, redundant processes, and fragmented authority. Understanding this labyrinth is critical to the Department of Government Efficiency's mission to streamline operations, eliminate waste, and enhance accountability.

At its core, the federal bureaucracy is an ecosystem comprising countless agencies, departments, and offices, each with its own mandate, budget, and operational framework. While this structure was originally intended to ensure that no single entity could dominate the governance process, it has led to significant challenges in coordination and communication. Agencies often operate in silos, pursuing narrowly defined objectives that may conflict with or duplicate the efforts of others. This lack of cohesion not only wastes resources but also undermines the government's ability to respond effectively to complex, cross-cutting issues.

The sheer scale of the federal government further complicates efforts to identify and address inefficiencies. From sprawling departments like Health and Human Services to specialized agencies such as NASA, the diversity of functions performed by federal entities reflects the breadth of the nation's needs. However, this diversity also creates opportunities for inefficiency to thrive, as responsibilities are divided and accountability becomes diffuse. Mapping this vast network requires a comprehensive approach that goes beyond surface-level assessments to examine the underlying structures and processes that drive government operations.

Data plays a pivotal role in this endeavor. By leveraging advanced analytics and machine learning, DOGE is able to dissect the complexities of the federal system, identifying patterns of inefficiency and pinpointing areas where reforms could have the greatest impact. This technological approach provides a level of insight that was previously unattainable, enabling the department to make data-driven decisions that are both precise and effective. For example, algorithms can analyze procurement processes to identify bottlenecks or highlight instances of redundant spending across multiple agencies.

In addition to technological tools, mapping the federal labyrinth requires a deep understanding of the human elements that sustain it. The culture of bureaucracy—shaped by decades of tradition, regulation, and political influence—plays a significant role in perpetuating inefficiencies. DOGE recognizes that any effort to reform the system must address not only structural issues but also the attitudes and behaviors of the people who work within it. This includes fostering a culture of collaboration and innovation, where public servants are empowered to think creatively and work across organizational boundaries.

Public engagement is another critical component of this mapping process. Citizens, as the ultimate stakeholders in government operations, possess valuable perspectives on how inefficiencies manifest in their daily lives. By incorporating their input through surveys, forums, and digital platforms, DOGE can ensure that its efforts are grounded in the realities of those it serves. This participatory approach not only enhances the accuracy of the mapping process but also builds public trust in the department's mission.

Ultimately, mapping the federal labyrinth is about more than cataloging inefficiencies; it is about creating a foundation for meaningful change. By understanding the intricacies of the system, DOGE can develop targeted strategies that address root causes rather than symptoms. This comprehensive approach ensures that reforms are not only impactful but also sustainable, laying the groundwork for a government that is more cohesive, efficient, and responsive to the needs of its citizens. Through this process, DOGE reaffirms its commitment to transforming the complexities of governance into a model of accountability and effectiveness.

Identifying Inefficiencies

Identifying inefficiencies within the federal government is a monumental task, requiring not only a comprehensive understanding of bureaucratic systems but also the innovative application of data-driven tools. At the heart of the Department of Government Efficiency's mission is the determination to expose and address the redundancies, misallocations, and outdated practices that have long hampered effective governance. These inefficiencies are more than operational flaws; they are barriers to the government's ability to fulfill its core mission of serving the public with transparency and accountability.

The first step in identifying inefficiencies is mapping the intricate web of federal agencies, programs, and initiatives that constitute the government's operational framework. Each layer of bureaucracy, while created to address specific challenges, often overlaps with others, leading to duplication of efforts and wasted resources. For instance, multiple agencies tasked with similar responsibilities, such as environmental regulation or public health, may operate in silos, unaware of or unwilling to coordinate their activities. This fragmentation not only inflates costs but also diminishes the effectiveness of service delivery, leaving citizens underserved and frustrated.

To tackle these challenges, DOGE employs advanced analytics and artificial intelligence to uncover patterns of inefficiency. By analyzing vast datasets from federal operations, AI can identify redundancies, highlight areas where resources are underutilized, and pinpoint bottlenecks that delay critical processes. This technological approach transforms what was once a painstaking manual review into a dynamic and actionable analysis, enabling the department to address inefficiencies with precision and speed.

One of the most striking examples of inefficiency lies in government procurement processes. Federal agencies collectively spend billions annually on goods and services, yet inefficiencies in procurement can lead to cost overruns, delays, and suboptimal outcomes. For example, the use of outdated contracting methods and the lack of coordination between agencies often result in unnecessary expenditures. DOGE's approach to reforming procurement includes standardizing processes across agencies and introducing real-time monitoring systems that ensure transparency and accountability in spending.

In addition to operational inefficiencies, the government faces significant challenges in human resource management. Many agencies are overstaffed in some areas while critically understaffed in others, reflecting a misalignment of workforce planning with actual needs. This imbalance not only strains budgets but also hampers the ability of agencies to fulfill their missions. Through workforce analytics, DOGE is identifying opportunities to optimize staffing levels, ensuring that human resources are allocated where they are most needed and can have the greatest impact.

Cultural factors within the bureaucracy also contribute to inefficiencies. Risk aversion, a hallmark of many government agencies, often stifles innovation and leads to a preference for maintaining the status quo. This mindset can result in the continuation of outdated practices, even when more effective alternatives are available. Addressing these cultural barriers requires fostering an environment where public servants are encouraged to think creatively and embrace change. By providing training and incentives for innovation, DOGE is working to shift the bureaucratic culture toward one that values efficiency and adaptability.

Ultimately, the process of identifying inefficiencies is not an end in itself but a means to achieving a government that is more responsive, effective, and aligned with the needs of its citizens. By combining cutting-edge technology with a deep understanding of systemic challenges, DOGE is laying the foundation for reforms that address the root causes of inefficiency. This transformative effort represents a crucial step toward rebuilding public trust in government institutions and ensuring that they are equipped to meet the demands of a rapidly changing world.

Agency Overlaps and Redundancies

Agency overlaps and redundancies are among the most glaring manifestations of inefficiency in the federal government. These issues arise when multiple entities are tasked with similar or overlapping responsibilities, leading to confusion, wasted resources, and diminished effectiveness. This duplication is not merely an administrative inconvenience; it is a systemic flaw that undermines the ability of the government to serve its citizens effectively and equitably.

At the heart of this problem lies the historical evolution of the federal bureaucracy. Over decades, new agencies, programs, and initiatives have been created to address emerging challenges or policy priorities. While each was established with a clear mandate, the cumulative effect has been the proliferation of entities with intersecting or identical functions. For example, multiple agencies are involved in disaster response, workforce development, and environmental protection, often working independently and without coordination. This fragmentation results in conflicting regulations, redundant processes, and inefficiencies that delay outcomes and inflate costs.

The consequences of these overlaps extend far beyond operational inefficiencies. For citizens, navigating the labyrinth of government services can be an overwhelming and frustrating experience. Businesses and organizations seeking permits, grants, or regulatory approvals are frequently required to interact with multiple agencies, each with its own requirements and timelines. This complexity not only hinders economic activity but also erodes public trust in the government's ability to function cohesively.

The Department of Government Efficiency has made addressing agency overlaps and redundancies a top priority. Utilizing advanced data analytics and organizational mapping tools, DOGE conducts comprehensive audits to identify areas where duplication occurs. These analyses go beyond surface-level assessments, examining the underlying processes and frameworks that perpetuate redundancy. By understanding these root causes, the department can propose targeted solutions that streamline operations without compromising the quality or accessibility of services.

A key component of this effort is fostering interagency collaboration. By breaking down silos and encouraging information-sharing, DOGE aims to create a culture of cooperation that minimizes duplication and enhances efficiency. For instance, establishing centralized databases and shared platforms enables agencies to coordinate their efforts and reduce redundant data collection or reporting requirements. These measures not only improve efficiency but also promote consistency and transparency in service delivery.

Consolidation is another strategy employed by DOGE to address overlaps. In cases where multiple agencies perform nearly identical functions, merging their operations can lead to significant cost savings and improved outcomes. Such consolidations are carefully planned to preserve the strengths of each agency while eliminating redundancies. For example, the integration of similar workforce training programs under a single administrative umbrella could streamline funding, standardize curricula, and expand access for participants.

While the benefits of addressing agency overlaps are clear, the process is not without challenges. Institutional resistance, political considerations, and the complexity of realigning existing structures can impede progress. To navigate these obstacles, DOGE emphasizes stakeholder engagement and transparent communication. By involving agency leaders, employees, and the public in the reform process, the department seeks to build consensus and ensure that changes are implemented thoughtfully and inclusively.

The elimination of redundancies is not merely an exercise in cost-cutting; it is a critical step toward building a government that is responsive, accountable, and equipped to meet the needs of its citizens. By addressing these inefficiencies, DOGE is helping to transform the federal bureaucracy into a cohesive and effective system, laying the foundation for a government that truly serves the public interest.

The Role of Data and AI

The role of data and artificial intelligence in reimagining government efficiency cannot be overstated. As the Department of Government Efficiency embarks on its transformative mission, these technologies serve as both the foundation and the catalyst for its initiatives. By leveraging data and AI, DOGE aims to unravel the complexities of the federal bureaucracy, illuminate inefficiencies, and introduce solutions that are both innovative and scalable.

Data is the lifeblood of modern governance. Within the federal government, it flows through every program, policy, and initiative, offering insights into how resources are allocated, where redundancies occur, and what outcomes are achieved. Historically, this data has often been underutilized, siloed within agencies, or obscured by outdated reporting mechanisms. DOGE's vision transforms this fragmented landscape into an integrated system where data is not merely collected but actively analyzed and acted upon. This shift from passive to proactive data management enables a level of precision in decision-making that was previously unattainable.

Artificial intelligence amplifies the power of data by introducing advanced analytics and predictive capabilities. Through machine learning algorithms, AI can identify patterns and anomalies across vast datasets, pinpointing inefficiencies that might otherwise go unnoticed. For instance, AI tools can analyze procurement records to detect wasteful spending or scrutinize regulatory processes to identify bottlenecks that delay approvals. This ability to process and interpret data at scale not only accelerates the identification of inefficiencies but also ensures that solutions are grounded in empirical evidence.

One of the most promising applications of AI within DOGE is in fraud detection. Federal programs disburse billions of dollars annually, making them attractive targets for fraud and abuse. Traditional methods of monitoring and enforcement often struggle to keep pace with the sophistication of fraudulent schemes. AI changes this dynamic by continuously scanning transactions, identifying irregularities, and flagging potential cases for investigation. This proactive approach not only saves money but also enhances public confidence in the integrity of government operations.

The integration of data and AI also extends to predictive modeling, enabling DOGE to anticipate future challenges and allocate resources more effectively. For example, by analyzing demographic trends, economic indicators, and historical data, AI systems can forecast demands on public services such as healthcare or infrastructure. This foresight allows the government to plan proactively, avoiding the reactive decision-making that has long characterized bureaucratic responses to crises.

However, the adoption of data and AI is not without its challenges. Ensuring data accuracy and integrity is paramount, as decisions based on flawed information can exacerbate rather than resolve inefficiencies. Moreover, the ethical implications of AI deployment, including concerns about bias, privacy, and accountability, must be addressed with rigor and transparency. DOGE recognizes these risks and has implemented robust safeguards, including data audits, algorithmic oversight, and stakeholder engagement, to ensure that its technological tools are used responsibly.

Transparency is another cornerstone of DOGE's approach to data and AI. By making its findings and methodologies accessible to the public, the department fosters trust and accountability. Citizens can see not only what decisions are being made but also the data and analysis that inform those decisions. This openness transforms data and AI from abstract concepts into tangible tools for public empowerment, demonstrating their potential to enhance, rather than undermine, democratic governance.

Through the strategic use of data and AI, DOGE is redefining what is possible in government efficiency. These technologies provide the means to break free from the constraints of traditional bureaucracy, introducing a new paradigm where decisions are smarter, operations are leaner, and outcomes are more aligned with public needs. In doing so, DOGE not only addresses the inefficiencies of the past but also sets a precedent for the future of governance, where innovation and accountability go hand in hand.

Harnessing Analytics for Reform

Harnessing analytics for reform within the Department of Government Efficiency represents a monumental leap toward transforming the federal bureaucracy. Analytics serves as the engine driving evidence-based decisions, enabling policymakers to navigate the labyrinth of inefficiencies with precision and clarity. It is not merely a tool for observation but a dynamic mechanism that transforms raw data into actionable insights, providing the foundation for meaningful and lasting reform.

At the heart of this approach lies the ability to aggregate and analyze data from across the vast network of federal agencies. Historically, these agencies have operated in silos, with minimal coordination or data sharing, leading to fragmented decision-making and missed opportunities for efficiency. By centralizing and integrating these data streams, analytics provides a panoramic view of government operations, highlighting redundancies, inefficiencies, and areas ripe for optimization. For example, a comprehensive analysis of procurement processes across departments might reveal overlapping contracts or opportunities to leverage collective bargaining for cost savings.

Predictive analytics takes this capability a step further by enabling foresight into future challenges and opportunities. Through the use of algorithms and statistical models, predictive tools can anticipate trends such as population growth, shifts in economic conditions, or emerging public health risks. This allows policymakers to allocate resources proactively, mitigating risks before they materialize. In the realm of infrastructure, predictive analytics could identify areas likely to experience critical failures, ensuring that maintenance funds are directed where they are most needed.

The implementation of analytics within DOGE also fosters accountability by establishing measurable benchmarks for success. Data-driven performance metrics allow the department to evaluate the effectiveness of reforms in real-time, ensuring that initiatives remain aligned with their objectives. For instance, if a program aimed at reducing administrative costs achieves only marginal savings, analytics can pinpoint the root causes of its underperformance, enabling adjustments to be made swiftly and effectively.

A notable application of analytics within DOGE is its role in identifying fraud, waste, and abuse. Federal programs disburse vast sums of money annually, making them attractive targets for exploitation. Advanced analytics tools can detect anomalies in spending patterns, flagging suspicious activities that warrant further investigation. This proactive approach not only recovers misappropriated funds but also deters future misconduct, reinforcing the integrity of government operations.

Despite its transformative potential, the use of analytics in governance is not without challenges. Ensuring the accuracy and reliability of data is paramount, as flawed inputs can lead to misguided policies. Equally important is addressing the ethical implications of data use, including concerns about privacy and the risk of reinforcing biases present in the underlying datasets. DOGE has established rigorous protocols to address these concerns, including data validation processes and transparent oversight mechanisms that ensure analytics serves the public interest without compromising individual rights.

The democratization of analytics also plays a crucial role in fostering public trust and engagement. By making key insights and findings accessible to citizens, DOGE transforms analytics from a technical tool into a bridge between government and the people it serves. Public dashboards and interactive platforms allow individuals to explore data relevant to their communities, providing a tangible demonstration of how reforms are progressing and where further work is needed.

In harnessing the power of analytics, DOGE is not only addressing the inefficiencies of the past but also laying the groundwork for a government that is agile, accountable, and equipped to navigate the complexities of the future. This approach exemplifies the potential of technology to drive systemic change, turning the abstract promise of reform into a concrete reality that benefits all citizens. Through analytics, the path to a more efficient and equitable government becomes not only visible but achievable, setting a new standard for what governance can and should be.

AI Tools for Fraud Detection

AI tools for fraud detection represent one of the most promising applications of technology in government reform, particularly within the ambitious framework of the Department of Government Efficiency. Fraud, waste, and abuse have long been persistent challenges in federal programs, siphoning resources from critical public services and undermining public trust. The integration of AI into fraud detection processes is a transformative step, offering unprecedented capabilities to identify, prevent, and mitigate fraudulent activities.

Fraud within government programs is not a new phenomenon. It manifests across a wide array of activities, from false claims in healthcare programs to misappropriations in federal contracts. The complexity and scale of federal operations have historically made fraud detection a daunting task, reliant on labor-intensive audits and whistleblower reports. These traditional methods, while valuable, are reactive and often detect fraud only after significant losses have occurred. AI changes this paradigm by enabling a proactive and predictive approach to fraud prevention.

The strength of AI in this domain lies in its ability to analyze vast volumes of data in real time. Machine learning algorithms can sift through millions of transactions, cross-referencing patterns, and identifying anomalies that human analysts might overlook. These anomalies often serve as early indicators of fraudulent activity, such as repeated billing for services not rendered or discrepancies between reported and actual outcomes in government-funded projects. By flagging these irregularities promptly, AI allows investigators to intervene before fraud escalates into significant financial or reputational damage.

Another critical advantage of AI tools is their adaptability. Traditional fraud detection systems are typically rule-based, relying on predefined criteria that can be bypassed by sophisticated schemes. In contrast, machine learning models evolve over time, learning from new data to refine their detection capabilities. This dynamic nature enables AI to stay ahead of emerging fraud techniques, continuously improving its effectiveness in an ever-changing landscape.

For example, in federal healthcare programs like Medicare and Medicaid, fraudulent claims have long been a significant issue. AI systems can analyze billing patterns to detect anomalies, such as a provider submitting an unusually high volume of claims for a rare procedure. Similarly, in procurement processes, AI can identify instances where contract awards deviate from established norms, flagging potential collusion or bid-rigging schemes. These tools not only enhance the accuracy of fraud detection but also reduce the time and resources required to uncover illicit activities.

Transparency and accountability are essential when deploying AI for fraud detection. Concerns about privacy and the potential for false positives necessitate robust governance frameworks to oversee the use of AI tools. The Department of Government Efficiency has prioritized these considerations, implementing safeguards to ensure that AI-driven investigations are fair, unbiased, and respectful of individual rights. Regular audits of algorithms and the inclusion of human oversight in decision-making processes help to mitigate risks and build public confidence in the system.

Public engagement also plays a pivotal role in the success of AI-driven fraud detection. By involving citizens in monitoring and reporting suspected fraud, DOGE enhances the efficacy of its initiatives while fostering a sense of shared responsibility. Crowdsourcing platforms and whistleblower protections encourage individuals to contribute to the fight against fraud, complementing the technological capabilities of AI systems.

Ultimately, the integration of AI tools for fraud detection within the Department of Government Efficiency exemplifies the potential of technology to address longstanding challenges in governance. By combining the speed and precision of AI with the expertise of human investigators, DOGE is setting a new standard for integrity and accountability in federal operations. This approach not only safeguards public resources but also reinforces the principles of trust and transparency that are fundamental to a functioning democracy. Through these innovations, the department is paving the way for a government that is not only efficient but also ethical and responsive to the needs of its citizens.

Regulatory Rollbacks

Regulatory rollbacks represent a critical component of the Department of Government Efficiency's strategy to streamline governance and enhance economic performance. Over decades, the federal regulatory framework has expanded to encompass a vast array of rules and mandates, each addressing specific societal concerns. While many of these regulations serve essential purposes, the cumulative effect has been to create a labyrinthine system that stifles innovation, burdens businesses, and complicates compliance for citizens and organizations alike.

At its core, the proliferation of regulations reflects a well-intentioned effort to protect public interests, ranging from environmental conservation to consumer safety. However, as new rules were layered onto existing ones, the system became increasingly unwieldy. Many regulations, though relevant at their inception, have outlived their usefulness or become redundant as industries and technologies evolved. Others conflict with newer policies, creating a web of contradictions that impedes effective enforcement and compliance. The result is a regulatory framework that often fails to achieve its intended goals while imposing significant costs on the economy.

For businesses, the regulatory burden manifests in the form of complex compliance requirements that drain resources and hinder growth. Small businesses, in particular, struggle to navigate the intricate maze of federal mandates, lacking the legal and administrative support that larger corporations can afford. This disparity not only stifles entrepreneurship but also exacerbates economic inequalities, as smaller entities are disproportionately impacted by regulatory inefficiencies. The cost of compliance is ultimately passed on to consumers, inflating prices and limiting access to goods and services.

Recognizing these challenges, the Department of Government Efficiency has prioritized the systematic review and rollback of outdated, redundant, and overly burdensome regulations. This initiative is not about deregulation for its own sake but about striking a balance between necessary oversight and operational efficiency. The goal is to eliminate rules that no longer serve the public interest while preserving those that provide essential protections. By doing so, DOGE aims to foster a regulatory environment that is both streamlined and responsive to the needs of a modern society.

The process begins with a comprehensive audit of existing regulations, leveraging advanced data analytics and stakeholder input to identify candidates for reform. This audit goes beyond surface-level assessments, examining the real-world impact of regulations on businesses, consumers, and government agencies. By analyzing patterns of compliance, enforcement, and outcomes, DOGE can pinpoint inefficiencies and prioritize areas for intervention. For instance, overlapping environmental regulations across multiple agencies might be consolidated into a single, coherent framework, reducing redundancy while maintaining environmental protections.

Engagement with stakeholders is a cornerstone of this effort. By involving businesses, advocacy groups, and citizens in the review process, DOGE ensures that reforms are informed by diverse perspectives and grounded in practical realities. Public comment periods, industry roundtables, and digital platforms provide opportunities for individuals and organizations to contribute their insights, fostering a sense of shared ownership in the reform process. This participatory approach not only enhances the quality of regulatory rollbacks but also builds trust in the government's commitment to serving the public good.

Transparency is equally vital in implementing regulatory reforms. DOGE is committed to making the rationale and outcomes of its actions accessible to the public, ensuring accountability at every stage. Real-time updates on regulatory reviews and rollbacks, available through online dashboards, allow citizens to track progress and understand how changes will affect them. This openness transforms a potentially contentious process into one of collaboration and trust-building.

The economic impact of regulatory rollbacks extends beyond immediate cost savings. By reducing administrative barriers and fostering a more business-friendly environment, these reforms encourage investment, innovation, and job creation. Industries previously hampered by regulatory inefficiencies can redirect resources toward growth and development, contributing to a more dynamic and competitive economy. At the same time, streamlined regulations enhance the government's capacity to enforce essential protections effectively, ensuring that oversight is both efficient and impactful.

Ultimately, regulatory rollbacks are about more than cutting red tape; they represent a reimagining of governance for the twenty-first century. By aligning regulations with current needs and capabilities, DOGE is creating a framework that supports innovation, accountability, and resilience. This approach not only addresses the inefficiencies of the past but also sets the stage for a government that is agile, transparent, and equipped to navigate the complexities of the future. Through these efforts, the Department of Government Efficiency is reaffirming its commitment to serving the public interest while fostering a vibrant and inclusive economy.

Simplifying Federal Codes

Simplifying federal codes is a critical step in addressing the inefficiencies that have long plagued the U.S. government. The existing labyrinth of laws, regulations, and administrative codes reflects decades of incremental additions, many of which have become redundant or outdated. While these codes were initially crafted to serve specific purposes, their accumulation has created a legal and regulatory environment that is often contradictory, inaccessible, and unmanageable. For both government officials and citizens, navigating this complex system has become an exercise in frustration, hindering effective governance and public trust.

The sheer volume of federal codes presents a daunting challenge. Over time, successive administrations and Congresses have layered new regulations onto the existing framework without systematically reviewing or retiring obsolete provisions. This approach has led to overlaps, inconsistencies, and unnecessary complexities. For example, businesses seeking permits or compliance certifications often encounter conflicting requirements from multiple agencies, each interpreting federal codes differently.

Such inefficiencies not only slow down decision-making but also inflate costs, diverting resources from innovation and growth.

For the average citizen, the complexity of federal codes can create significant barriers to accessing government services or understanding their legal obligations. Whether it is navigating the tax code, applying for federal benefits, or complying with regulatory requirements, the lack of clarity in federal codes often leaves individuals overwhelmed and disadvantaged. This systemic opacity disproportionately affects marginalized communities, who may lack the resources or expertise to navigate such a convoluted system.

Recognizing the urgent need for reform, the Department of Government Efficiency has made simplifying federal codes a central focus of its mission. This initiative is not merely about reducing the volume of regulations but about creating a streamlined and coherent legal framework that enhances accessibility, accountability, and effectiveness. The process begins with a comprehensive audit of existing codes to identify redundancies, contradictions, and provisions that no longer serve their intended purpose. By leveraging advanced data analytics and natural language processing tools, DOGE can systematically evaluate the relevance and impact of each regulation, ensuring that decisions are data-driven and transparent.

Eliminating unnecessary codes is only part of the solution. DOGE also aims to rewrite and reorganize the remaining regulations to ensure clarity and coherence. This involves standardizing language, consolidating overlapping provisions, and restructuring codes into a format that is intuitive and user-friendly. For instance, digital platforms can be developed to provide searchable databases of federal codes, allowing users to quickly locate relevant regulations and understand their implications. These tools empower citizens and businesses alike, fostering greater compliance and engagement with government policies.

A crucial aspect of this reform is stakeholder involvement. DOGE recognizes that the success of simplifying federal codes depends on input from those directly impacted by these regulations. Through public consultations, workshops, and online forums, the department seeks to gather insights and feedback from businesses, advocacy groups, legal experts, and ordinary citizens. This collaborative approach not only enhances the quality of the reforms but also builds public trust in the process.

Simplifying federal codes also has significant economic implications. Streamlined regulations reduce compliance costs for businesses, encouraging investment and innovation. For the government, clear and efficient codes translate into faster decision-making and reduced administrative burdens, freeing up resources for critical priorities. Furthermore, a simplified legal framework strengthens the rule of law by making it easier for citizens to understand and adhere to their obligations, reducing unintentional violations and disputes.

The broader impact of simplifying federal codes extends to the very fabric of democracy. By creating a legal system that is transparent, accessible, and equitable, DOGE aims to restore public confidence in government institutions. This initiative is not just about operational efficiency but about reaffirming the government's commitment to serving its people with integrity and fairness. In doing so, the Department of Government Efficiency is laying the groundwork for a governance model that is both modern and resilient, capable of meeting the challenges of the twenty-first century with clarity and purpose.

Evaluating the Economic Impact

Evaluating the economic impact of regulatory reform is central to understanding the transformative potential of the Department of Government Efficiency's initiatives. Regulations, while often necessary to protect public interests, have significant implications for economic activity. They shape business practices, influence investment decisions, and affect the cost structures of industries. As such, any effort to streamline or simplify regulations must be accompanied by a careful assessment of its economic consequences.

Regulations impose costs on businesses and individuals, both directly and indirectly. Direct costs include compliance expenses such as reporting requirements, licensing fees, and modifications to business operations to meet regulatory standards. Indirect costs, often less visible but equally burdensome, stem from the delays and inefficiencies that complex regulations introduce into economic processes. For example, the time spent navigating bureaucratic requirements represents an opportunity cost, diverting resources from innovation and productivity.

These costs disproportionately affect small and medium-sized enterprises, which lack the financial and administrative resources to manage complex regulatory landscapes. Larger corporations often have the capacity to absorb these costs or navigate them more efficiently, creating an uneven playing field that stifles competition and innovation. The economic impact is not limited to businesses; consumers ultimately bear the burden through higher prices and reduced access to goods and services.

Streamlining federal regulations has the potential to unlock significant economic benefits. By reducing unnecessary complexity and eliminating redundancies, regulatory reform can lower compliance costs, enhance efficiency, and stimulate entrepreneurial activity. Businesses freed from excessive regulatory burdens are more likely to invest in growth, adopt new technologies, and expand their operations, creating jobs and driving economic development.

The positive ripple effects extend beyond businesses. Simplified regulations can enhance government efficiency by reducing administrative overhead and enabling agencies to focus on their core missions. This, in turn, can improve service delivery, benefitting citizens and fostering greater public trust in government institutions. Moreover, streamlined regulations can enhance the competitiveness of the U.S. economy on the global stage by reducing the costs of doing business and attracting foreign investment.

To evaluate the economic impact of regulatory reform, the Department of Government Efficiency employs advanced data analytics and economic modeling tools. These technologies enable a detailed analysis of how changes to specific regulations affect economic indicators such as employment, productivity, and gross domestic product. For instance, a cost-benefit analysis might quantify the savings generated by eliminating a redundant reporting requirement while considering the potential risks or trade-offs associated with the reform.

Stakeholder engagement is also a critical component of this evaluation process. By consulting with businesses, industry associations, consumer advocacy groups, and other stakeholders, DOGE ensures that its reforms are informed by a diverse range of perspectives and grounded in real-world considerations. This collaborative approach not only enhances the quality of regulatory changes but also builds consensus and support for the reform agenda.

Transparency and accountability are essential in evaluating the economic impact of regulatory reform. DOGE is committed to sharing its findings with the public, providing clear and accessible explanations of how specific changes are expected to benefit the economy. This openness fosters trust and allows citizens to see the tangible outcomes of reform efforts, reinforcing their confidence in the government's ability to act in the public interest.

While the potential economic benefits of regulatory reform are significant, they must be balanced against the need to maintain essential protections for health, safety, and the environment. The goal is not to deregulate indiscriminately but to ensure that regulations achieve their intended purposes without imposing unnecessary costs or hindering economic progress. By striking this balance, DOGE is working to create a regulatory framework that supports both public welfare and economic vitality.

The economic impact of regulatory reform extends beyond immediate cost savings; it represents a broader transformation in how the government interacts with the economy. By fostering a more efficient, transparent, and responsive regulatory environment, DOGE is paving the way for a government that enables, rather than impedes, economic growth and innovation. This approach not only addresses the inefficiencies of the present but also lays the foundation for a more prosperous and resilient future.

Chapter 3: Participatory Governance

A Government for the People

A government truly for the people represents more than a democratic ideal; it embodies the promise of a governance system that prioritizes transparency, accountability, and responsiveness. Central to this vision is the Department of Government Efficiency, an institution committed to redefining the relationship between the federal government and its citizens. The department's mission underscores the imperative to build systems that not only function effectively but also reflect the values and aspirations of the people they serve.

Historically, government structures have struggled to balance the demands of administrative efficiency with the diverse needs of a populous nation. This challenge has been compounded by the complexities of modern governance, where layers of bureaucracy often obscure the government's ultimate purpose: serving its citizens. The result has been a growing disconnect, marked by public disillusionment and a perceived erosion of trust in governmental institutions. Addressing this disconnect requires a fundamental rethinking of how the government operates and how it engages with its constituents.

DOGE's approach to creating a government for the people begins with a commitment to transparency. A transparent government is one that opens its processes, decisions, and data to public scrutiny, fostering an environment where accountability is not just an ideal but a practical reality. Transparency ensures that citizens have a clear understanding of how their tax dollars are spent, how decisions are made, and how policies impact their daily lives. Through initiatives like public accountability dashboards and open-data platforms, DOGE seeks to demystify the workings of government, transforming it from an opaque entity into a participatory space where citizens can engage, question, and influence.

Equally critical is the emphasis on efficiency without sacrificing integrity. A government for the people must manage its resources judiciously, eliminating waste and redundancies while preserving the programs and services that are vital to public welfare. DOGE's innovative use of technology, particularly artificial intelligence and advanced analytics, enables the department to identify inefficiencies and optimize processes. For example, streamlining agency operations not only reduces costs but also accelerates service delivery, ensuring that citizens receive timely and effective support.

Participation lies at the heart of a government designed for the people. Recognizing that citizens are not merely passive recipients of governance but active contributors, DOGE has pioneered initiatives to incorporate public input into decision-making processes. Crowdsourcing platforms, virtual town halls, and community engagement forums provide avenues for individuals to voice their ideas, concerns, and solutions. This participatory model not only enriches policy development but also strengthens the social contract between the government and its people, fostering a sense of shared responsibility and collective purpose.

At its core, the vision of a government for the people is a vision of empowerment. It seeks to equip citizens with the tools and knowledge needed to navigate and shape the systems that affect their lives. Whether through simplified processes, accessible information, or opportunities for direct involvement, DOGE aims to break down the barriers that have historically alienated individuals from their government. This approach recognizes that true efficiency extends beyond operational metrics; it is measured by the extent to which government actions resonate with and respond to the needs of its people.

The journey toward a government for the people is not without challenges. It requires confronting entrenched practices, overcoming resistance to change, and navigating the complexities of modern governance. However, the rewards of this transformation are profound. A government that is transparent, efficient, and participatory not only functions more effectively but also inspires trust and pride among its citizens. It becomes a living embodiment of democratic ideals, a system in which governance is not imposed but co-created.

Through its bold initiatives, DOGE is redefining what it means to govern in the twenty-first century. It is demonstrating that a government for the people is not an abstract aspiration but a tangible reality, achievable through innovation, collaboration, and an unwavering commitment to public service. As the department continues its work, it is setting a precedent for governance that is not only effective but also deeply aligned with the principles of equity, justice, and inclusivity. This is the promise of a government for the people: a system that serves, empowers, and uplifts all who depend on it.

Democratizing Efficiency

Democratizing efficiency is a cornerstone of the vision that underpins the Department of Government Efficiency. This concept goes beyond streamlining operations or reducing redundancies; it reflects a fundamental shift in the governance paradigm, one that places citizens at the heart of reform initiatives. By redefining efficiency as a shared endeavor rather than a top-down mandate, DOGE aims to transform public administration into a collaborative process that aligns with the values and aspirations of the people it serves.

At its essence, democratizing efficiency involves breaking down the traditional barriers that have long separated citizens from decision-making processes. Historically, governance has often been perceived as an opaque and inaccessible institution, where policies are crafted in isolation from the lived realities of those they affect. This detachment has not only fostered distrust but has also undermined the effectiveness of government programs. DOGE seeks to reverse this trend by creating systems that invite participation, transparency, and accountability, ensuring that efficiency becomes a collective goal rather than an imposed directive.

One of the most transformative aspects of democratizing efficiency is the integration of participatory tools that empower citizens to contribute directly to governance. Digital platforms and crowd-sourced initiatives allow individuals to voice their ideas, report inefficiencies, and propose solutions. These mechanisms transform the public from passive recipients of government services into active participants in shaping policy and reform. For instance, the creation of an open-access database of federal expenditures enables citizens to track how funds are allocated and spent, fostering a culture of transparency and shared accountability.

The democratization of efficiency also extends to fostering equity in how reforms are implemented. Traditional efficiency measures often focus narrowly on cost-cutting or resource optimization, sometimes at the expense of marginalized communities. DOGE's approach is fundamentally different, emphasizing that efficiency must not come at the cost of equity or inclusivity. By involving diverse stakeholders in decision-making processes, the department ensures that reforms are not only effective but also just. This commitment to inclusivity reflects an understanding that a government cannot be efficient if it fails to meet the needs of all its citizens.

Technology plays a pivotal role in enabling this shift toward democratized efficiency. Advances in artificial intelligence and data analytics provide the tools needed to identify inefficiencies and optimize processes at an unprecedented scale. Yet, the application of these technologies is guided by a principle that human oversight and participation remain integral to the process. AI-driven insights must complement, not replace, the nuanced understanding that comes from lived experiences and community engagement. This balance between technological innovation and human input ensures that efficiency reforms remain grounded in the realities of those they aim to serve.

A critical challenge in democratizing efficiency lies in overcoming resistance to change, both within government institutions and among the public. Bureaucracies, by their nature, are often resistant to reform, particularly when it involves relinquishing control or altering established workflows. Similarly, citizens may approach participatory initiatives with skepticism, shaped by a history of unfulfilled promises or limited transparency. Addressing these challenges requires a sustained commitment to communication, education, and trust-building. DOGE's outreach efforts, including public forums, educational campaigns, and transparent reporting, are designed to bridge these gaps and foster a sense of shared purpose.

The ultimate goal of democratizing efficiency is to create a governance system that reflects and responds to the collective intelligence of its people. By embedding efficiency within a participatory framework, DOGE is redefining what it means to govern in a democratic society. It is not merely about doing more with less but about ensuring that every action, decision, and resource allocation serves the public good in a way that is transparent, equitable, and inclusive. This vision of democratized efficiency represents a profound transformation in governance, one that promises to rebuild trust, enhance accountability, and set a new standard for public administration in the twenty-first century.

Crowdsourcing Solutions

Crowdsourcing solutions within governance represents an innovative and transformative approach to solving complex societal challenges. This strategy leverages the collective intelligence, creativity, and expertise of citizens, positioning them as active participants in shaping government policies and operations. For the Department of Government Efficiency, crowdsourcing serves not only as a method for generating actionable ideas but also as a vital mechanism for rebuilding public trust and fostering a more inclusive democracy.

The essence of crowdsourcing lies in its ability to tap into diverse perspectives. Traditional government decision-making often occurs within insular frameworks, where policymakers operate in isolation from the broader populace. This detachment can lead to policies that, while well-intentioned, fail to address the nuanced realities of those they aim to serve. Crowdsourcing disrupts this paradigm by inviting citizens, industry experts, and community leaders to contribute their insights and solutions. By doing so, it democratizes the process of governance, ensuring that decisions are informed by the lived experiences and expertise of a wide array of stakeholders.

Technological advancements have made crowdsourcing more accessible and effective than ever before. Digital platforms now enable large-scale collaboration, allowing governments to solicit input from millions of individuals in real time. For example, DOGE might implement an interactive online portal where citizens can propose ideas for eliminating waste, improving service delivery, or streamlining bureaucratic processes. Users could upvote or comment on proposals, creating a dynamic forum where the most viable and popular ideas rise to prominence. This approach not only generates innovative solutions but also fosters a sense of ownership and engagement among participants.

The benefits of crowdsourcing extend beyond idea generation. It also serves as a powerful tool for identifying inefficiencies and holding government accountable. Citizens often possess unique insights into the shortcomings of government services, whether through their experiences with public agencies or their observations of waste and redundancy. By creating channels for this feedback, DOGE can gain valuable intelligence on areas requiring reform while demonstrating its commitment to transparency and responsiveness.

Crowdsourcing also aligns with the principles of agility and adaptability that underpin DOGE's mission. In an era of rapid technological and societal change, traditional policymaking processes can struggle to keep pace. Crowdsourcing enables governments to respond more swiftly to emerging challenges by drawing on the collective ingenuity of the public. Whether addressing a sudden crisis or exploring long-term reforms, this approach ensures that policies remain relevant and effective in a constantly evolving landscape.

Despite its potential, implementing crowdsourcing within governance is not without challenges. Ensuring the quality and relevance of contributions requires robust mechanisms for vetting and evaluating submissions. Additionally, the process must be inclusive, addressing barriers such as digital divides that may prevent certain groups from participating. DOGE's commitment to equity demands that it actively work to ensure that all voices, particularly those from marginalized communities, are heard and valued in the crowdsourcing process.

Trust is another critical factor. For crowdsourcing to succeed, citizens must believe that their contributions will be taken seriously and lead to tangible outcomes. This requires a transparent and accountable framework for evaluating, implementing, and reporting on ideas generated through the process. By providing regular updates on the status of crowdsourced initiatives and showcasing successful implementations, DOGE can build and sustain public confidence in its efforts.

The integration of crowdsourcing into the fabric of governance marks a profound shift in how governments interact with their citizens. It transforms governance from a top-down directive into a collaborative partnership, where power and responsibility are shared in the pursuit of common goals. For DOGE, this approach represents a cornerstone of its vision for a more efficient, inclusive, and responsive government. By harnessing the collective wisdom of the nation, it is not only solving immediate challenges but also laying the foundation for a governance model that embodies the democratic ideals of participation and accountability.

Transparency in Action

Transparency in governance is more than a principle; it is a practical imperative that fosters trust, accountability, and efficacy in public administration. Within the Department of Government Efficiency, transparency is not treated as an ancillary goal but as a foundational value driving all reform initiatives. By opening the workings of government to public scrutiny, the department aims to dismantle the walls of opacity that have historically separated policymakers from the people they serve, ensuring that accountability is woven into the fabric of every decision and action.

Achieving meaningful transparency requires moving beyond the traditional measures of publishing reports or conducting audits. While these efforts remain essential, they are often retrospective and fail to provide the real-time clarity necessary for a modern, participatory democracy. Transparency in action, as envisioned by DOGE, embraces the transformative potential of digital tools and data analytics to create a dynamic system where information flows freely, empowering citizens and enhancing oversight.

The first step in fostering transparency lies in making government operations comprehensible and accessible to all. Public accountability dashboards, for instance, offer an innovative solution by presenting key metrics—such as budget allocations, project progress, and performance outcomes—in an intuitive, user-friendly format. These platforms allow citizens to track the status of government initiatives, offering insights into how taxpayer dollars are spent and whether objectives are being met. By eliminating technical jargon and prioritizing clarity, these tools ensure that transparency is not limited to those with specialized knowledge but is a resource for every citizen.

Transparency also demands proactive communication. Citizens should not have to dig through dense bureaucratic archives to understand the actions of their government. Regular updates through digital platforms, public forums, and media channels create a continuous dialogue between the government and the public. These efforts not only inform citizens but also invite their feedback, fostering a culture of shared responsibility and collaboration. For DOGE, transparency is inherently interactive, requiring the government to listen as much as it speaks.

A cornerstone of transparency is the accessibility of data. Open-data initiatives, championed by DOGE, ensure that raw government data is made available to the public, researchers, and private entities. These datasets fuel innovation, enabling independent analysis and the development of applications that can enhance public understanding and engagement. Whether tracking environmental trends, analyzing healthcare expenditures, or identifying patterns of inefficiency, open data equips citizens with the tools needed to hold their government accountable.

The technological foundation supporting transparency is critical. Advances in artificial intelligence and machine learning offer unparalleled opportunities to analyze vast amounts of information quickly and accurately. These technologies enable the identification of inefficiencies, inconsistencies, or even potential corruption within the government. For example, AI algorithms can detect anomalies in procurement patterns, flagging irregularities that warrant investigation. By leveraging these tools, DOGE not only ensures internal accountability but also demonstrates its commitment to ethical and responsible governance.

However, transparency is not without its challenges. Balancing openness with security and privacy is a delicate task. Certain government operations, particularly those involving national security or sensitive personal data, require confidentiality. DOGE's approach to transparency acknowledges these limitations while striving to maximize openness wherever possible. Clear policies delineating what information can be shared and why ensure that transparency does not compromise safety or privacy.

The cultural shift required to embed transparency within governance cannot be overstated. For decades, government institutions have often prioritized control over information as a means of maintaining authority. Reversing this mindset involves redefining transparency as a strength rather than a vulnerability. DOGE's leadership exemplifies this ethos, modeling openness in their own practices and setting expectations for the agencies under their purview. This cultural transformation, while gradual, is essential for institutionalizing transparency across all levels of government.

Transparency in action is not merely a strategy for reform; it is a reinvigoration of democracy itself. By illuminating the inner workings of governance, DOGE reaffirms the fundamental contract between the government and its citizens—a contract based on trust, accountability, and mutual respect. In doing so, it sets a new standard for what modern governance can and should be: open, inclusive, and driven by the collective will and wisdom of the people it serves. This vision of transparency is not only an ideal but a practical necessity, ensuring that government remains a force for good in the lives of its citizens.

The Suggestion Box Initiative

The Suggestion Box Initiative represents a profound shift in how the government interacts with its citizens, redefining the dynamics of governance through direct public engagement. By creating a platform where individuals can contribute their insights and ideas, the Department of Government Efficiency transforms governance from a static hierarchy into a dynamic, participatory process. This initiative exemplifies DOGE's commitment to democratizing efficiency and fostering transparency, ensuring that governance becomes a collective endeavor.

At its core, the Suggestion Box Initiative is a digital forum designed to harness the collective intelligence of the nation. It invites citizens, public servants, and stakeholders from all sectors to propose innovative solutions, highlight inefficiencies, and suggest reforms. This platform is not merely a repository for ideas; it is a curated space where contributions are evaluated, refined, and, where feasible, implemented. By democratizing access to decision-making processes, the initiative ensures that governance reflects the diversity and ingenuity of the populace it serves.

The effectiveness of the Suggestion Box Initiative lies in its accessibility and inclusivity. Designed to accommodate a wide range of participants, the platform eliminates barriers to entry, ensuring that every citizen has a voice. It employs user-friendly interfaces and multilingual support to reach underserved communities, empowering individuals who might otherwise be excluded from policy discussions. This approach not only enriches the quality of submissions but also fosters a sense of ownership and engagement among participants, reinforcing the social contract between the government and its citizens.

Technology is the backbone of this initiative. Advanced algorithms categorize and analyze submissions, identifying recurring themes and prioritizing proposals based on feasibility, impact, and public support. Machine learning models are employed to flag innovative ideas, detect patterns in the data, and ensure that no valuable contribution is overlooked. This integration of technology streamlines the evaluation process, enabling DOGE to process large volumes of submissions efficiently while maintaining a high standard of review.

Transparency is integral to the success of the Suggestion Box Initiative. Participants are provided with real-time updates on the status of their submissions, from initial review to implementation. This feedback loop not only validates their contributions but also builds trust in the process. By publicly showcasing successful proposals and acknowledging their contributors, DOGE highlights the tangible impact of citizen engagement, encouraging further participation and fostering a culture of innovation.

The initiative also serves as a powerful tool for uncovering inefficiencies and redundancies within government operations. Citizens often encounter bureaucratic hurdles and systemic inefficiencies firsthand, making them uniquely positioned to identify areas in need of reform. By providing a platform for these insights, the Suggestion Box Initiative transforms individual experiences into actionable intelligence, bridging the gap between lived realities and policy decisions.

While the potential of the Suggestion Box Initiative is immense, its success depends on addressing several challenges. Ensuring the quality and relevance of submissions requires robust moderation and curation processes. Equally important is the need to safeguard the platform from misuse, such as spamming or the submission of false information. DOGE addresses these challenges through a combination of human oversight and technological safeguards, maintaining the integrity and credibility of the initiative.

The broader impact of the Suggestion Box Initiative extends beyond the ideas it generates. It represents a cultural shift toward participatory governance, where citizens are no longer passive recipients of policies but active contributors to their formulation. This transformation is not only empowering but also instrumental in restoring public trust in government institutions. By demonstrating that citizen input is valued and impactful, the initiative fosters a renewed sense of agency and optimism among the populace.

In essence, the Suggestion Box Initiative encapsulates DOGE's vision of a government that is responsive, inclusive, and innovative. It leverages the collective intelligence of the nation to address its most pressing challenges, creating a model of governance that is not only efficient but also deeply aligned with the principles of democracy. As the initiative evolves, it has the potential to redefine the relationship between the government and its citizens, setting a precedent for participatory governance in the twenty-first century.

Public Accountability Dashboards

Public Accountability Dashboards embody a revolutionary step in reimagining the relationship between government institutions and the citizens they serve. These platforms are not merely tools for disseminating information; they represent a profound commitment to transparency, participation, and accountability. By transforming raw data into accessible and actionable insights, these dashboards bridge the gap between governance and the public, ensuring that government actions are both visible and comprehensible to those who entrust it with authority.

The concept of a public accountability dashboard is rooted in the belief that citizens have the right to understand and evaluate the performance of their government. Traditional methods of transparency, such as annual reports and press releases, often fail to engage the public due to their complexity or inaccessibility. In contrast, dashboards leverage modern technology to present critical information in a user-friendly and interactive format. Through these platforms, citizens can track key metrics such as budget allocations, project timelines, and service delivery outcomes, gaining real-time insights into the workings of their government.

A cornerstone of these dashboards is their ability to distill complex governmental operations into clear and digestible visuals. Advanced data visualization tools transform vast datasets into charts, graphs, and maps that convey trends, progress, and areas of concern at a glance. This approach not only enhances understanding but also encourages engagement, empowering citizens to ask informed questions and provide constructive feedback. For the Department of Government Efficiency, these dashboards serve as both a window into its activities and a mechanism for fostering dialogue with the public.

The impact of public accountability dashboards extends beyond transparency; they also act as catalysts for efficiency and innovation. By exposing inefficiencies and inconsistencies, these platforms incentivize government agencies to improve their performance. Departments are no longer evaluated solely by internal metrics but are held accountable to public scrutiny. This external perspective often highlights issues that may be overlooked within traditional bureaucratic processes, driving continuous improvement and fostering a culture of accountability.

A defining feature of these dashboards is their emphasis on inclusivity. Recognizing that accessibility is crucial to their effectiveness, DOGE has prioritized the development of platforms that cater to diverse audiences. Multilingual interfaces, mobile compatibility, and features for individuals with disabilities ensure that no segment of the population is excluded from participating in this new era of transparency. This commitment to inclusivity reflects DOGE's broader vision of a government that serves all its citizens equitably and effectively.

Technology plays a pivotal role in the operation of public accountability dashboards. Integrating artificial intelligence and machine learning, these platforms can analyze trends, identify anomalies, and predict future outcomes. For instance, an AI-driven feature might flag discrepancies in project expenditures or highlight regions where service delivery lags behind targets. These insights enable proactive decision-making, allowing DOGE to address issues before they escalate into crises. Moreover, by automating routine analysis, these technologies free up human resources for more strategic tasks, further enhancing operational efficiency.

Despite their transformative potential, the implementation of public accountability dashboards is not without challenges. Ensuring the accuracy and security of the data presented is paramount. Any errors or breaches could undermine public trust and compromise the initiative's credibility. DOGE has addressed these concerns by establishing rigorous protocols for data verification and cybersecurity, demonstrating its commitment to maintaining the integrity of its platforms.

The cultural shift required to fully leverage public accountability dashboards cannot be understated. For decades, government agencies have operated with limited external oversight, often perceiving transparency as a threat rather than an opportunity.

Transitioning to a system where performance is openly evaluated by the public requires not only technological investment but also a reorientation of institutional mindsets. DOGE's leadership has been instrumental in navigating this transition, emphasizing that transparency strengthens rather than diminishes governmental authority.

In the broader context of governance, public accountability dashboards symbolize a redefinition of the social contract. They reaffirm the principle that government exists to serve its people, not in isolation but in partnership. By providing citizens with the tools to hold their leaders accountable, these platforms empower individuals to become active participants in democracy. This shift from passive observation to active engagement represents a profound transformation in the way governance is conceived and practiced.

Ultimately, public accountability dashboards are more than a technological innovation; they are a statement of values. They reflect a commitment to openness, equity, and collaboration, setting a new standard for how governments interact with their citizens. As DOGE continues to refine and expand these platforms, it charts a path toward a future where governance is not only efficient but also deeply responsive to the needs and aspirations of the people it serves.

Engaging the Citizenry

Engaging the citizenry is not a peripheral objective but a fundamental principle of the Department of Government Efficiency. At its core, this initiative recognizes that governance is most effective when it is inclusive, participatory, and responsive to the voices of the people it serves. The evolving dynamics of modern democracy demand more than passive representation; they call for active engagement that bridges the gap between government institutions and the citizenry. Through innovative strategies and platforms, DOGE seeks to foster a renewed sense of civic involvement, ensuring that every individual feels empowered to contribute to the governance process.

The essence of engaging the citizenry lies in fostering a culture of dialogue and collaboration. Citizens are not merely recipients of government services but are integral stakeholders in the decision-making process. This philosophy is operationalized through town halls, community forums, and digital platforms that create spaces for open communication. These channels allow for the exchange of ideas, concerns, and feedback, enabling policymakers to craft solutions that are grounded in the realities of those they aim to serve.

Digital transformation plays a pivotal role in enhancing civic engagement. In an era where technology permeates every aspect of life, leveraging digital tools to connect with citizens is both a necessity and an opportunity. DOGE's approach includes the development of user-friendly online platforms where individuals can participate in polls, submit suggestions, and engage in meaningful discussions about policy initiatives. These platforms are designed to be accessible to all, ensuring that participation is not hindered by technological barriers or digital divides.

Transparency is a cornerstone of effective citizen engagement. By making information readily available and understandable, DOGE builds trust and encourages active participation. Public dashboards and interactive portals not only provide insights into government operations but also invite scrutiny and accountability. This openness fosters a sense of ownership among citizens, who see their feedback and contributions reflected in tangible outcomes.

The challenges of engaging the citizenry are not to be underestimated. Ensuring inclusivity requires deliberate efforts to reach marginalized and underserved communities. This involves addressing linguistic, cultural, and economic barriers that may hinder participation. DOGE's commitment to equity is evident in its strategies to bridge these gaps, from multilingual support to targeted outreach initiatives that bring the conversation to communities that have historically been excluded from the policy-making process.

Building trust is central to this endeavor. Decades of perceived inefficiency, corruption, and unresponsiveness have eroded public confidence in government institutions. Rebuilding this trust requires consistent and genuine efforts to demonstrate that citizen input is valued and impactful. DOGE achieves this through transparency in action—providing regular updates on policy progress, showcasing successful collaborations, and acknowledging the contributions of individuals and communities.

Civic education is another vital component of this initiative. Empowering citizens to engage effectively requires equipping them with the knowledge and tools to navigate the complexities of governance. Educational campaigns, workshops, and online resources are integral to DOGE's strategy, demystifying government operations and fostering a more informed and active citizenry.

The ultimate goal of engaging the citizenry is to create a governance model that is truly of the people, by the people, and for the people. This vision transcends the traditional boundaries of government-citizen interactions, envisioning a collaborative partnership where power and responsibility are shared. For DOGE, this approach is not merely an ideal but a practical necessity, ensuring that governance remains adaptive, inclusive, and aligned with the evolving needs of society.

Through its emphasis on engagement, DOGE redefines the social contract between the government and its citizens. It transforms governance from a hierarchical structure into a participatory network, where every voice matters, and every contribution counts. This approach not only enhances the effectiveness of government initiatives but also revitalizes the democratic spirit, fostering a sense of unity and purpose that is essential for the nation's progress. By placing the citizen at the heart of its mission, DOGE paves the way for a government that truly serves its people with integrity, transparency, and accountability.

Town Halls and Online Platforms

Town halls and online platforms represent the dual pillars of a reimagined public engagement strategy that blends tradition with innovation. At its core, this approach seeks to bridge the divide between government and citizens, fostering a dynamic dialogue that reflects the diversity and complexity of modern governance.

By revitalizing the timeless institution of the town hall while embracing the transformative potential of digital platforms, DOGE advances a participatory model of governance where every voice has the opportunity to shape the policies that define a nation.

The town hall meeting has long been a cornerstone of American democracy, providing a forum for citizens to connect directly with their leaders. DOGE's vision breathes new life into this tradition, ensuring its relevance in the context of twenty-first-century challenges. Modern town halls extend beyond physical gatherings, incorporating hybrid models that allow participation both in-person and virtually. This inclusivity ensures that geography, mobility, or other barriers no longer hinder engagement, opening the doors for widespread participation.

These events are meticulously designed to be more than venues for passive listening. They serve as interactive dialogues where citizens can raise concerns, propose ideas, and engage with policymakers in real-time. Structured formats prioritize equitable participation, ensuring that a broad spectrum of perspectives is heard. Facilitators skilled in mediation and discourse guide these interactions, maintaining focus and fostering respectful exchanges, even on contentious topics.

Complementing the town hall is the expansive reach of online platforms, which transcend the limitations of physical presence. These platforms, developed with user accessibility and inclusivity in mind, are tailored to engage a digitally connected populace. Interactive features such as live Q&A sessions, polls, and forums provide citizens with multiple avenues to express their views and contribute to policy discussions. For DOGE, these platforms are not just tools for communication but vehicles for co-creation, where citizens and policymakers collaborate to refine initiatives and solve problems.

The true power of online platforms lies in their capacity to amplify voices that might otherwise go unheard. Multilingual support, user-friendly interfaces, and compatibility across devices ensure that these platforms are accessible to all, including marginalized communities and those with limited technological proficiency. This democratization of access is central to DOGE's ethos, reflecting its commitment to an inclusive approach to governance.

Technology also facilitates the aggregation and analysis of citizen input, transforming disparate voices into actionable insights. Machine learning algorithms categorize and prioritize feedback, identifying trends and uncovering emerging issues that may not yet be on policymakers' radar. This data-driven approach ensures that public engagement informs decision-making at every level, from the allocation of resources to the formulation of national strategies.

Transparency and accountability underpin both town halls and online platforms. For citizens, knowing that their input matters and seeing tangible outcomes from their contributions builds trust and sustains engagement. DOGE's commitment to these principles is evident in its emphasis on providing regular updates and feedback loops, where participants can track the progress of initiatives they helped shape.

However, the success of these efforts is contingent on overcoming several challenges. Ensuring the authenticity of interactions, particularly on digital platforms, requires robust systems to prevent misuse, such as misinformation or attempts to game the system. Additionally, fostering meaningful dialogue in large-scale town halls or highly active online platforms demands careful moderation to maintain focus and civility. DOGE addresses these issues through a combination of advanced technology and human oversight, ensuring that engagement remains productive and constructive.

The synergy between town halls and online platforms embodies a governance model that is both deeply rooted in tradition and boldly innovative. Together, they create a participatory ecosystem where citizens are not only informed but empowered to shape the trajectory of their nation. In doing so, DOGE redefines public engagement as a cornerstone of effective governance, setting a precedent for how democratic institutions can evolve to meet the needs of their people.

Building Trust Through Communication

Building trust through communication is fundamental to the success of the Department of Government Efficiency, as trust serves as the cornerstone of any functional relationship between government and citizens. In an era marked by widespread skepticism and diminished confidence in public institutions, the need for transparent, authentic, and inclusive communication has never been more urgent. For DOGE, this principle is not merely aspirational; it is an operational necessity that underpins its mission to reform governance while fostering genuine public engagement.

Effective communication begins with transparency, a value that is often professed but seldom realized in the intricate workings of bureaucracy. DOGE commits to laying bare the processes, decisions, and outcomes of its initiatives, offering an unfiltered view of its operations. This approach demystifies the inner workings of governance, replacing suspicion and doubt with clarity and understanding. Citizens are no longer passive recipients of decisions but informed participants who can assess the rationale behind policies and track their implementation.

To achieve this level of transparency, DOGE employs a multifaceted strategy that leverages both traditional and digital communication channels. Town halls, press briefings, and community forums provide direct access to policymakers, fostering a culture of accountability. Meanwhile, digital platforms extend this accessibility to a broader audience, offering real-time updates and interactive features that invite public scrutiny and dialogue. These platforms are designed not only to disseminate information but to facilitate two-way communication, where citizens can voice their concerns and contribute their insights.

Authenticity is another critical component of trust-building. In an age where public discourse is often marred by spin and misinformation, the ability to communicate honestly and without pretense sets DOGE apart. By acknowledging challenges, admitting mistakes, and demonstrating a willingness to adapt, the department cultivates a relationship with citizens rooted in mutual respect and shared purpose. Authentic communication fosters resilience, ensuring that trust endures even in the face of setbacks or controversies.

Equally important is the inclusivity of DOGE's communication efforts. Recognizing the diversity of the American populace, the department prioritizes accessibility in all its initiatives. Multilingual resources, culturally sensitive messaging, and accommodations for individuals with disabilities ensure that no voice is excluded. This inclusivity extends to the design of its platforms and tools, which are crafted to be intuitive and user-friendly, empowering citizens of all backgrounds to engage meaningfully with their government.

Feedback loops are a vital mechanism in this framework, transforming communication from a one-sided broadcast into a collaborative exchange. By actively soliciting input and acting on it, DOGE demonstrates that citizen perspectives are not only heard but valued. Regular updates on the status of reforms, including successes, challenges, and adjustments, provide tangible evidence of this commitment. These feedback loops create a virtuous cycle of trust, where transparency leads to engagement, and engagement reinforces accountability.

The broader implications of this approach extend beyond the immediate objectives of DOGE. By fostering trust through effective communication, the department lays the groundwork for a more engaged and empowered citizenry. This transformation has the potential to revitalize democratic governance, shifting the paradigm from one of detachment and disillusionment to one of collaboration and shared responsibility. In this model, communication is not merely a tool but a bridge that connects the aspirations of the people with the actions of their government.

Ultimately, the trust cultivated through these efforts is not an end in itself but a means to achieve a higher purpose. It enables DOGE to navigate the complexities of reform, rally public support, and overcome resistance from entrenched interests. Trust empowers citizens to believe in the possibility of change and to contribute to its realization. For DOGE, this trust is both the foundation and the catalyst for building a government that truly serves its people with integrity, efficiency, and accountability.

Chapter 4: Fiscal Responsibility in Action

Cutting the Fat

Eliminating waste in government operations represents one of the most pressing imperatives for the Department of Government Efficiency. Bureaucratic inefficiencies have long been a hallmark of the federal system, manifesting in redundant programs, outdated technologies, and a labyrinthine structure resistant to change. For decades, these inefficiencies have siphoned away critical resources, weakening the effectiveness of government services and undermining public trust. The call to "cut the fat" is not merely about cost reduction; it is a moral and practical necessity to realign government spending with the values of transparency, accountability, and results-driven performance.

At the core of this initiative is the identification and eradication of programs and processes that no longer serve their intended purpose or have become outdated in an era of rapid technological advancement. This requires a comprehensive audit of federal agencies, examining budgets, operational procedures, and program outcomes with a level of scrutiny and precision that has been historically absent. By deploying advanced data analytics and machine learning, DOGE can identify inefficiencies that might otherwise remain hidden, uncovering patterns of wasteful spending and revealing opportunities for significant cost savings.

One of the most egregious forms of waste arises from duplicative programs, where multiple agencies or departments perform overlapping functions without coordination. These redundancies not only inflate costs but also create confusion and inefficiencies that ripple through the system. Addressing this issue requires not just streamlining operations but fostering interagency collaboration to ensure that resources are allocated where they are most needed. In this context, consolidation of efforts becomes a tool for enhancing efficiency without compromising service quality.

Another area ripe for reform is the reliance on antiquated technologies that hinder rather than help. Many government systems, particularly those managing critical data or public services, operate on software and hardware that have long exceeded their life expectancy. Maintenance costs for these legacy systems are exorbitant, often surpassing the cost of implementing modern alternatives. Yet, institutional inertia and a resistance to change have perpetuated their existence. DOGE's approach involves not only replacing outdated systems but ensuring that technological upgrades align with a long-term strategy for sustainability and scalability.

The cultural dimensions of waste within government cannot be overlooked. Decades of entrenched practices have created a system where inefficiency is often normalized, and the fear of disruption outweighs the potential benefits of reform. Changing this culture is as critical as addressing the structural and technological issues. DOGE's leadership emphasizes a shift toward a results-oriented mindset, rewarding innovation and risk-taking within a framework of accountability. Civil servants are encouraged to identify inefficiencies and propose solutions, fostering a collaborative environment where reform is a shared responsibility.

The process of cutting waste must also be balanced with sensitivity to its potential human impact. Workforce reductions and program eliminations, while necessary in some cases, must be approached with care to avoid unintended consequences. This includes providing support and retraining for displaced workers, ensuring that reforms do not exacerbate socio-economic disparities. The goal is not simply to reduce costs but to redirect resources toward initiatives that deliver measurable benefits to the public, creating a government that is both lean and compassionate.

Ultimately, the effort to eliminate waste is about restoring integrity and purpose to federal operations. Each dollar saved through efficient management is a dollar that can be reinvested in education, healthcare, infrastructure, and other vital services. It is about demonstrating to the American people that their government is capable of functioning effectively and responsibly, rebuilding the trust that has been eroded by decades of mismanagement. Through its unwavering commitment to cutting the fat, DOGE seeks to lay the foundation for a federal system that embodies the principles of efficiency, equity, and excellence.

Identifying and Eliminating Wasteful Programs

Identifying and eliminating wasteful programs requires a meticulous and unbiased examination of federal operations, a process that blends data-driven methodologies with a commitment to public accountability. Waste in governance is not simply a matter of excessive spending; it reflects deeper structural issues, including misaligned priorities, bureaucratic inertia, and a lack of rigorous oversight. Addressing these inefficiencies demands both precision and the political will to make difficult, yet necessary, decisions.

The initial step in tackling wasteful programs involves a comprehensive audit of all federal agencies. This audit goes beyond a superficial review of budgets and expenditures, delving into the specific outcomes of each program to assess their relevance, effectiveness, and alignment with national goals. Programs that fail to meet their objectives or have become obsolete in the face of technological or societal advancements are prime candidates for reevaluation. This rigorous assessment relies on transparent criteria to ensure that decisions are grounded in evidence rather than political expediency.

Advanced technologies play a pivotal role in identifying inefficiencies. Machine learning algorithms and predictive analytics provide insights into spending patterns and operational redundancies that are invisible to traditional auditing methods. By analyzing vast datasets, these tools can pinpoint anomalies, such as disproportionate administrative costs or overlapping functions across agencies, allowing policymakers to focus their reform efforts where they will have the greatest impact.

One of the most challenging aspects of eliminating waste is addressing redundancies across multiple agencies. Over decades, federal operations have evolved into a sprawling network where similar tasks are often duplicated by different departments. For instance, overlapping environmental monitoring programs or redundant grant application processes can waste both financial resources and personnel hours. Streamlining these functions requires consolidating efforts under a unified framework, a process that must be executed carefully to maintain service quality and avoid disruption.

The process also entails confronting entrenched interests that resist change. Programs, once established, often develop their own ecosystems of stakeholders, including contractors, advocacy groups, and even legislative supporters, all of whom may oppose their elimination regardless of inefficiency. Successfully navigating this resistance involves clear communication about the rationale for reforms, coupled with measures to mitigate the impact on affected stakeholders. For example, reallocating resources from discontinued programs to initiatives with demonstrated success can help ease the transition.

Equally important is the human element of waste reduction. Programs often represent livelihoods for federal employees, contractors, and local communities. Addressing waste without accounting for these human costs risks alienating the public and undermining the broader objectives of reform. Workforce transitions, retraining programs, and community support initiatives are essential components of an ethical approach to downsizing. By prioritizing these measures, DOGE ensures that reforms are both fiscally responsible and socially conscientious.

Eliminating wasteful programs is not an isolated endeavor but part of a larger commitment to building a government that is lean, effective, and aligned with public priorities. Each program that is restructured, consolidated, or discontinued represents a step toward restoring public trust and ensuring that taxpayer dollars are used wisely. This process requires continuous vigilance and a willingness to adapt as new challenges emerge, embodying the ethos of a government that is truly efficient and accountable. Through these efforts, DOGE aims to redefine the standards of federal governance, proving that efficiency and integrity are not mutually exclusive but mutually reinforcing.

Managing the Federal Workforce

Managing the federal workforce is both a delicate art and a critical component of achieving government efficiency. A streamlined, adaptable, and motivated workforce is essential to the success of the Department of Government Efficiency. The challenge lies not only in optimizing the size and structure of the federal workforce but also in fostering a culture of innovation, accountability, and service excellence within its ranks. This undertaking requires balancing fiscal responsibility with the ethical imperatives of maintaining livelihoods and preserving institutional knowledge.

The federal workforce has grown over decades, often in response to emerging needs or crises, resulting in a sprawling network of employees, contractors, and support staff spread across numerous agencies. While these expansions were often necessary in their time, they have collectively contributed to inefficiencies that hamper the system today. Redundant roles, outdated skill sets, and a lack of interagency coordination are common symptoms of a workforce that has not evolved in step with technological advancements and societal demands. Reforming this system begins with a comprehensive assessment of its composition and performance.

Data-driven approaches play a pivotal role in understanding workforce dynamics. Advanced analytics can reveal inefficiencies in staffing patterns, identify skill gaps, and assess productivity at both the individual and agency levels. These insights provide the foundation for informed decisions about workforce restructuring, allowing the government to allocate human resources where they are most needed. This process is not about indiscriminately reducing headcounts but about aligning the workforce with the strategic goals of the federal system.

At the heart of workforce reform is the need to modernize job roles and skill requirements. As automation and artificial intelligence continue to transform the nature of work, the federal government must adapt by reskilling its employees for tasks that require human ingenuity, judgment, and empathy. Programs to upskill workers in data analysis, digital communication, and strategic planning ensure that the workforce remains relevant and capable of addressing contemporary challenges. These initiatives also demonstrate a commitment to employee development, fostering loyalty and engagement.

Workforce management must also address the cultural dimensions of reform. Many federal employees operate within rigid hierarchies that discourage risk-taking and innovation. Transforming this culture requires leadership that prioritizes creativity, flexibility, and accountability. By creating environments where employees feel empowered to propose solutions and take ownership of their work, the government can tap into the vast potential of its workforce. Recognition and rewards for exceptional performance further reinforce this shift, establishing a merit-based culture that values results over adherence to outdated protocols.

While restructuring efforts may necessitate workforce reductions, these changes must be approached with care to minimize social and economic disruptions. Transition programs that include severance packages, career counseling, and job placement services help ease the impact on affected employees. Moreover, targeted support for regions or communities disproportionately reliant on federal employment ensures that reforms do not exacerbate inequality or economic instability.

Managing the federal workforce is a dynamic and ongoing process that requires agility and foresight. It is not enough to implement changes and hope they endure; continuous evaluation and adaptation are necessary to respond to evolving needs and technologies. By fostering a workforce that is lean, skilled, and motivated, the federal government can not only achieve operational efficiency but also set a standard for public service excellence. This transformation is central to the mission of DOGE, reaffirming its commitment to building a government that is both effective and humane.

Strategic Budgeting

Strategic budgeting is the lifeblood of any government, an essential mechanism that underpins the effective allocation of resources and the realization of policy goals. In an era where fiscal pressures mount and public expectations rise, the art of budgeting must transcend mere accounting to become a dynamic and anticipatory process. It is within this context that the Department of Government Efficiency redefines how federal funds are prioritized, allocated, and managed, transforming strategic budgeting into a cornerstone of governance.

The essence of strategic budgeting lies in its ability to balance immediate needs with long-term objectives, ensuring that each dollar spent contributes meaningfully to the nation's growth and stability. This approach requires a shift from reactive spending, which addresses crises as they arise, to proactive planning that anticipates future challenges and opportunities. By embracing data-driven methodologies, the federal government can craft budgets that are not only fiscally responsible but also strategically aligned with the nation's priorities.

Central to this evolution is the integration of advanced technologies. Artificial intelligence and predictive analytics offer unparalleled insights into economic trends, demographic shifts, and program performance. These tools enable policymakers to model various scenarios, assessing the potential impact of spending decisions on a wide range of outcomes. By simulating the long-term effects of budgetary choices, the government can identify the most effective pathways to achieve its goals while minimizing unintended consequences.

Transparency is another pillar of strategic budgeting. The allocation of public funds must be a process that invites scrutiny and fosters trust. Clear criteria for budgetary decisions, coupled with accessible reporting mechanisms, ensure that taxpayers understand how their money is being used. This transparency not only enhances accountability but also empowers citizens to engage with and influence the budgeting process, reinforcing the democratic principles at the heart of governance.

Efficiency is equally critical. Strategic budgeting demands a rigorous examination of expenditures to eliminate waste and optimize resource allocation. This involves identifying programs that deliver high returns on investment, as well as those that fail to achieve their intended outcomes. By reallocating funds from underperforming initiatives to more impactful ones, the government can maximize the benefits of its spending while addressing areas of greatest need.

However, the success of strategic budgeting depends on more than just technical proficiency; it requires political will and institutional commitment. The process must be insulated from short-term partisan interests that prioritize electoral gains over sustainable development. Achieving this requires a cultural shift within government agencies, where long-term planning is valued as highly as immediate results. Leadership plays a pivotal role in fostering this mindset, setting the tone for a budgeting process that is disciplined, innovative, and future-focused.

Strategic budgeting also has profound implications for interagency coordination. Federal agencies must collaborate to align their objectives and share resources effectively, breaking down silos that hinder efficiency. This collaborative approach ensures that budgetary decisions are informed by a holistic understanding of the nation's needs and opportunities, rather than being confined to the narrow perspectives of individual departments.

Ultimately, strategic budgeting is not merely a financial exercise but a declaration of priorities and values. It reflects a government's commitment to using its resources wisely, equitably, and sustainably, addressing the needs of today without compromising the prospects of tomorrow. Through its emphasis on innovation, accountability, and foresight, the Department of Government Efficiency sets a new standard for how the federal budget can serve as a tool for national transformation. In doing so, it reaffirms the promise of governance that is not only responsible but also profoundly attuned to the aspirations of its people.

AI for Predictive Budget Analysis

Artificial intelligence is revolutionizing the field of budgetary planning, offering unprecedented tools for predictive analysis and strategic foresight. Within the framework of the Department of Government Efficiency, AI-driven predictive budget analysis emerges as a linchpin for reimagining fiscal management. This technology transforms traditional budgeting into a dynamic process that anticipates challenges, optimizes resource allocation, and aligns expenditures with long-term national objectives.

The promise of predictive budget analysis lies in its ability to process vast amounts of data from diverse sources—economic trends, demographic shifts, historical spending patterns, and global market indicators. This information, when synthesized through sophisticated algorithms, enables a granular understanding of the potential impacts of budgetary decisions. Unlike static models of the past, AI-powered tools continuously update projections, reflecting real-time developments and allowing policymakers to respond with agility.

At its core, predictive budget analysis enhances decision-making by quantifying the trade-offs inherent in fiscal planning. Policymakers face complex questions: Should funds be diverted from existing programs to invest in new infrastructure? How will demographic changes influence healthcare expenditures? AI models simulate these scenarios, providing insights into their economic and social ramifications. This level of precision ensures that budgeting decisions are not only financially prudent but also aligned with broader policy goals, such as equity and sustainability.

AI's capacity to identify inefficiencies is another critical asset. By analyzing patterns of expenditure, predictive models can pinpoint areas where funds are underutilized or misallocated. For instance, an AI system might flag redundancies in overlapping federal programs or reveal disparities between allocated budgets and actual performance outcomes. Armed with this information, the Department of Government Efficiency can target reforms that maximize the impact of each taxpayer dollar.

Furthermore, predictive budget analysis strengthens transparency and accountability. By visualizing projections and outcomes through user-friendly interfaces, these tools enable stakeholders—lawmakers, citizens, and oversight bodies—to engage with the budgeting process more effectively. This openness fosters trust, as decisions are grounded in evidence and communicated with clarity. Public confidence in the government's fiscal stewardship is reinforced when citizens can trace how their contributions are translated into tangible societal benefits.

However, the integration of AI into budgetary planning is not without challenges. The quality of predictions hinges on the integrity and comprehensiveness of the data used. Incomplete or biased data sets can skew outcomes, leading to flawed decisions. Ensuring robust data governance—standardizing inputs, validating sources, and addressing potential biases—is therefore paramount. Additionally, the ethical implications of automating budgetary decisions must be carefully navigated, as overreliance on algorithms risks sidelining human judgment and contextual understanding.

To address these challenges, the Department of Government Efficiency must foster collaboration between technologists, economists, and policymakers. By aligning technical expertise with domain-specific knowledge, the government can harness AI's full potential while safeguarding against its pitfalls. Training programs that equip decision-makers with the skills to interpret and act on AI-generated insights are equally essential.

The transformative power of predictive budget analysis lies not only in its capacity to streamline expenditures but also in its ability to elevate governance. It signals a shift toward a government that is proactive, adaptive, and deeply attuned to the needs of its constituents. By embedding AI at the heart of its fiscal strategy, the Department of Government Efficiency sets a new standard for how technology can drive meaningful reform. This approach not only addresses immediate fiscal challenges but also lays the foundation for a resilient, forward-looking public finance system.

Long-Term Fiscal Planning

Long-term fiscal planning is the cornerstone of sustainable governance, a process that demands foresight, adaptability, and a commitment to balancing present needs with future obligations. Within the framework of the Department of Government Efficiency, the integration of advanced methodologies and technological innovations into fiscal planning promises to revolutionize the way the federal government approaches resource allocation and budgetary priorities.

At its core, long-term fiscal planning seeks to address one of the government's most pressing challenges: the balance between economic stability and the public good. This requires a shift from short-term reactive budgeting to a proactive and holistic strategy. By focusing on forecasting and scenario analysis, policymakers can anticipate potential fiscal pressures, such as demographic shifts, technological advancements, and unforeseen economic disruptions. These insights allow for the development of resilient strategies that safeguard against financial crises while promoting growth and equity.

Artificial intelligence plays a pivotal role in this evolution. Through machine learning algorithms and predictive analytics, long-term fiscal planning can identify emerging trends and simulate the impact of various policy decisions. For instance, an AI model could analyze the economic ripple effects of investing in renewable energy infrastructure versus traditional sectors. By providing a data-driven framework, these tools enable policymakers to weigh trade-offs and make informed decisions that align with broader national objectives.

Incorporating these technologies also facilitates better resource prioritization. With finite financial resources, the federal government must determine which programs yield the highest return on investment, both economically and socially. Long-term fiscal planning provides a mechanism to evaluate the efficacy of public expenditures, distinguishing between essential services and areas of potential reallocation. This ensures that every dollar spent contributes to measurable outcomes, such as reduced inequality, improved public health, and enhanced infrastructure.

However, the success of long-term fiscal planning hinges on overcoming significant challenges. Chief among these is the need for bipartisan collaboration and institutional support. Fiscal sustainability is not a partisan issue but a shared responsibility that transcends electoral cycles. Achieving this requires building consensus among stakeholders, fostering a culture of accountability, and insulating budgetary processes from short-term political pressures.

Transparency is another critical component. Engaging the public in fiscal planning processes not only enhances trust but also enriches the decision-making framework with diverse perspectives. Tools such as public dashboards and interactive modeling platforms can demystify fiscal data, enabling citizens to understand and contribute to the planning process. By cultivating an informed electorate, the government can strengthen democratic participation and foster collective ownership of fiscal outcomes.

Moreover, long-term fiscal planning must account for global economic dynamics. The interconnectedness of modern economies means that domestic fiscal policies are inextricably linked to international trends, such as trade agreements, currency fluctuations, and geopolitical shifts. Integrating global considerations into fiscal strategies ensures that the United States remains competitive and resilient in an increasingly complex economic landscape.

The Department of Government Efficiency is uniquely positioned to lead this transformative effort. By championing innovative approaches and leveraging cutting-edge technology, it sets a precedent for how long-term fiscal planning can address systemic inefficiencies and build a foundation for sustainable prosperity. This vision is not merely an administrative exercise but a profound commitment to securing the financial health and well-being of future generations. Through foresight, collaboration, and a relentless focus on outcomes, long-term fiscal planning becomes more than a strategy—it becomes a legacy of responsible and effective governance.

Balancing Cost with Impact

Balancing cost with impact is a delicate art in governance, requiring a profound understanding of not only the financial metrics but also the broader implications of public spending on societal well-being. Within the framework of the Department of Government Efficiency, this balance transcends traditional budgetary considerations to embody a holistic approach that prioritizes outcomes over optics. It is not enough to cut costs;

every reduction must be weighed against its potential ripple effects on the fabric of society.

At the core of this endeavor lies the recognition that government expenditures are more than line items in a ledger. They represent investments in the nation's future, touching every aspect of citizens' lives, from education and healthcare to infrastructure and national security. The challenge, therefore, is to ensure that cost-cutting measures do not undermine the fundamental services that enable societal progress and equity. Balancing cost with impact means adopting a strategy that safeguards these essentials while eliminating inefficiencies that detract from their potential.

This strategy begins with an honest assessment of what works and what does not. Programs and initiatives must be evaluated not only for their financial outlays but also for their effectiveness in achieving intended objectives. Data analytics and performance metrics become indispensable tools in this process, allowing policymakers to identify areas where resources are misaligned or underutilized. By focusing on measurable outcomes, the government can redirect funds from underperforming programs to those that deliver tangible benefits, ensuring that every dollar spent creates value.

However, the pursuit of efficiency cannot come at the expense of equity. Reducing costs in areas that disproportionately affect vulnerable populations risks exacerbating social inequalities and eroding public trust. For example, cuts to healthcare programs or educational funding may yield immediate savings but can result in long-term societal costs, such as higher unemployment rates, poorer health outcomes, and increased reliance on social safety nets. Balancing cost with impact requires a commitment to protecting these critical services, even as the government seeks to streamline operations.

Moreover, this balancing act necessitates transparency and public engagement. Citizens must be informed about the rationale behind budgetary decisions and given a voice in the process. This participatory approach not only fosters trust but also ensures that policies reflect the diverse needs and priorities of the population. Mechanisms such as public forums, online consultations, and interactive dashboards can demystify fiscal policies, allowing citizens to see how their tax dollars are being used and to contribute to shaping budgetary priorities.

Innovation also plays a pivotal role in this process. By leveraging technology, the government can achieve efficiencies without compromising quality or accessibility. For instance, digital platforms can streamline service delivery, reducing administrative costs while enhancing user experience. Similarly, investments in renewable energy and smart infrastructure can yield long-term savings, reducing environmental impact and positioning the nation as a leader in sustainable development.

The Department of Government Efficiency embodies this vision, demonstrating that fiscal responsibility and societal impact are not mutually exclusive. By fostering a culture of accountability and innovation, it aims to set a precedent for how governments can manage resources effectively while upholding their obligations to the public. This approach is not merely about cutting costs but about reimagining governance to align with the principles of equity, transparency, and sustainability.

In balancing cost with impact, the government acknowledges its dual responsibility as a steward of public resources and a champion of the public good. It is a commitment to ensuring that every decision, every dollar, and every policy contributes to a vision of a nation that thrives not only economically but also socially and morally. Through careful planning and unwavering dedication to these principles, the Department of Government Efficiency aspires to transform this ideal into a tangible reality.

Avoiding Detrimental Cuts

Avoiding detrimental cuts in government spending is an exercise in foresight, balance, and prudence. It demands a keen understanding of the interconnected nature of federal programs and the cascading effects that reductions in one area can have across the broader societal and economic landscape. The challenge lies in distinguishing between cuts that eliminate waste and inefficiency and those that undermine the critical functions of governance, disrupt essential services, or disproportionately harm vulnerable populations.

Effective governance requires the recognition that not all expenditures are created equal. Some represent investments in long-term stability and growth, while others are symptomatic of inefficiency or redundancy. The key to avoiding detrimental cuts is a rigorous evaluation of outcomes and impacts. This process goes beyond traditional cost-benefit analyses to encompass a deeper assessment of social value. For example, while reducing funding for public education or healthcare may offer immediate fiscal savings, the long-term consequences—ranging from diminished economic productivity to higher social welfare costs—can far outweigh the initial benefits.

The process begins with identifying programs and services that are indispensable to the public good. These are areas where cuts would lead to significant disruptions in quality of life, public safety, or economic stability. For instance, investments in infrastructure, public health, and education often yield returns that extend far beyond their initial costs, contributing to national competitiveness and resilience. Similarly, programs targeting underserved communities play a critical role in reducing inequality and fostering social cohesion, making them essential to a balanced and inclusive society.

Data-driven decision-making becomes an invaluable tool in this context. By employing advanced analytics and performance metrics, policymakers can gain a nuanced understanding of which programs deliver the highest value relative to their costs. This approach allows for precision in budgetary adjustments, ensuring that reductions target inefficiencies rather than core functions. Machine learning and predictive modeling further enhance this capability, providing insights into the potential ripple effects of proposed cuts and enabling the design of mitigation strategies.

Avoiding detrimental cuts also requires robust stakeholder engagement. Policymakers must consult with affected communities, industry leaders, and subject matter experts to gain a comprehensive understanding of how budgetary changes will play out on the ground. Such consultations not only provide valuable perspectives but also foster trust and transparency in the decision-making process. Public forums, surveys, and collaborative workshops can serve as platforms for gathering input and building consensus around difficult choices.

Moreover, the government must explore alternative strategies to achieve fiscal goals without compromising critical services. This could include revenue-generating measures, such as tax reforms or public-private partnerships, as well as innovative approaches to cost reduction. For example, adopting digital technologies to streamline administrative processes or renegotiating contracts to secure better terms can lead to substantial savings without the need for service reductions.

The ethical dimension of budgetary decisions cannot be overstated. Cuts that disproportionately affect vulnerable populations—such as children, the elderly, or low-income families—risk deepening social divides and eroding public trust in government institutions. Policymakers must be guided by principles of equity and fairness, ensuring that the burden of fiscal adjustments is shared equitably and that the most disadvantaged are protected from undue harm.

The Department of Government Efficiency embodies this commitment to thoughtful and responsible fiscal management. Its approach is rooted in a vision of governance that prioritizes effectiveness and accountability while safeguarding the values of equity and inclusivity. By leveraging technology, engaging stakeholders, and adopting a holistic perspective, the department seeks to navigate the complexities of fiscal reform with integrity and precision.

In navigating the difficult terrain of budget cuts, the ultimate goal is not merely to reduce expenditures but to create a government that is both efficient and responsive to the needs of its citizens. This balance is critical to ensuring that fiscal responsibility does not come at the expense of societal progress or public well-being. By avoiding detrimental cuts, the government reaffirms its commitment to serving its people with prudence, compassion, and foresight, laying the foundation for a future defined by stability and opportunity.

Protecting Essential Services

The safeguarding of essential services during a government overhaul presents a paradoxical challenge: trimming inefficiency without severing lifelines. In any initiative seeking to restructure bureaucracy, particularly one as ambitious as the Department of Government Efficiency, the stakes of failing to protect critical public goods cannot be overstated. The Department's mission to eradicate waste must not jeopardize the social, economic, and ethical imperatives underpinning essential services like healthcare, education, and emergency response systems.

Fundamentally, essential services represent the connective tissue of a society, ensuring stability, security, and equitable access to opportunities. These services—designed to meet the needs of the most vulnerable—are not mere expenditures on a ledger but investments in the nation's human capital and social fabric. Yet, the temptation to streamline operations often risks undermining these very safeguards, particularly in a climate of fiscal austerity. This necessitates a nuanced approach, one that pairs fiscal prudence with moral responsibility.

The first step in achieving this balance lies in precision. Through advanced data analytics, DOGE has the capacity to delineate waste from necessity. By employing artificial intelligence tools to audit expenditures and model outcomes, the department can identify redundancies and optimize resource allocation without impairing service delivery. For instance, administrative bloat within a healthcare program can be reduced without diminishing the quality of patient care by reallocating savings toward frontline services or technological upgrades. This approach ensures that efficiencies do not come at the expense of efficacy.

Moreover, the prioritization of essential services must be informed by robust public engagement. Citizens, as beneficiaries of these services, possess invaluable insights into their functionality and areas for improvement. Town hall discussions, digital surveys, and participatory forums can help policymakers understand the societal impacts of potential cuts. This grassroots input, coupled with empirical evidence, creates a feedback loop that strengthens decision-making and minimizes unintended consequences.

Equally critical is the role of contingency planning. The unpredictable nature of crises—be it pandemics, natural disasters, or economic shocks—demands a government prepared to respond with agility and sufficiency. DOGE's reforms must incorporate resilience measures, ensuring that essential services can withstand fiscal tightening without compromising their capacity to act in times of need. Establishing emergency funds, streamlining procurement processes, and fostering inter-agency cooperation are all strategies that enhance the government's ability to safeguard public welfare even amid budgetary constraints.

Finally, maintaining the integrity of essential services requires a clear articulation of the values underpinning governance. Efficiency must never eclipse equity. The reshaping of federal operations must align with the fundamental ethos of serving the public good. This moral clarity not only guides reform efforts but also fosters trust among citizens, reinforcing their faith in the government's commitment to their well-being.

In preserving essential services, DOGE exemplifies the delicate art of reform: pursuing innovation without forsaking compassion. It embodies a model of governance where efficiency serves as a vehicle, not a barrier, to equity and social progress. By adopting strategies that are both bold and conscientious, the department charts a path forward that honors the dual imperatives of fiscal responsibility and human dignity.

Chapter 5: Ethical Dimensions of Efficiency

The Morality of Downsizing

The morality of downsizing is a complex subject, woven deeply into the ethical fabric of societal governance and public administration. As organizations and governments grapple with the challenges of balancing budgets, increasing efficiency, and fostering innovation, the human cost of these decisions cannot be overlooked. Downsizing, often couched in the language of streamlining or restructuring, is fraught with moral considerations that challenge leaders to weigh the benefits of efficiency against the societal impacts of job displacement.

The essence of the dilemma lies in reconciling the need for a leaner, more efficient government with the foundational principles of equity and dignity for those affected. Downsizing is not merely a technical process of trimming budgets or optimizing workflows; it is a human process that affects lives, communities, and the collective trust in governance. The decision to reduce workforce sizes, consolidate departments, or phase out roles must be approached with a profound sense of responsibility, ensuring that those impacted are not left behind in the pursuit of operational efficiency.

Central to this ethical quandary is the imperative to treat displaced workers with dignity and fairness. This includes providing robust support systems such as retraining programs, transitional assistance, and opportunities for redeployment within the system. By investing in the human capital affected by these changes, leaders affirm their commitment to the principles of inclusivity and social responsibility. It is not enough to simply implement changes for the sake of efficiency; these changes must reflect a government's dedication to the well-being of its people.

Another critical dimension is the broader societal impact of downsizing. Reductions in government jobs can reverberate through local economies, particularly in regions where federal employment forms a significant portion of the economic landscape. These cascading effects highlight the necessity for strategic planning and community engagement. Decisions must be made with an awareness of their ripple effects, balancing immediate fiscal benefits with long-term societal stability.

Moreover, the process of downsizing must be transparent and participatory. Engaging stakeholders, including employees, unions, and the public, ensures that the rationale for changes is clearly communicated and understood. This fosters trust and minimizes resistance by emphasizing that these measures are not arbitrary but necessary for the greater good. A transparent process also enables accountability, ensuring that downsizing efforts are implemented equitably and without favoritism or bias.

Ultimately, the morality of downsizing is not a binary question of right or wrong but a multifaceted challenge that demands thoughtful navigation. It requires leaders to prioritize humanity over expediency, ensuring that efficiency does not come at the expense of the social contract between government and its citizens. By embracing a holistic approach that balances fiscal responsibility with ethical stewardship, governments can demonstrate that downsizing, when done responsibly, can lead to a stronger, more resilient society.

Balancing Jobs and Efficiency

The balance between preserving jobs and ensuring government efficiency represents one of the most morally and practically complex issues in public administration. At its heart, this tension reflects a collision of values: the ethical responsibility to safeguard livelihoods and the pragmatic necessity of achieving a lean, effective bureaucracy. Navigating this terrain requires a nuanced understanding of the stakes, the context, and the solutions that honor both imperatives.

The federal workforce has historically been a bedrock of stable employment, offering millions of Americans the security of steady income, benefits, and pensions. It has also served as a pathway to upward mobility, especially for historically underserved populations. The notion of reducing this workforce, even in the name of efficiency, inevitably carries with it significant social and economic consequences. Downsizing risks not only displacing individuals but also unraveling the intricate social fabric woven around government employment. Families, communities, and local economies often depend heavily on the stability provided by federal jobs, particularly in regions where alternative employment opportunities are scarce.

However, the inefficiencies endemic to the federal bureaucracy also exact their toll, albeit in less visible ways. Redundant positions, outdated workflows, and unnecessary processes divert resources that could otherwise be directed toward critical public goods such as education, healthcare, and infrastructure. Every inefficiency represents an opportunity cost—a missed chance to invest in innovation, social equity, and economic growth. The imperative to address these inefficiencies is not about cutting for its own sake but about aligning resources with the nation's evolving priorities and needs.

The challenge, then, is to balance these competing demands: to create a government that is both efficient and humane. A key aspect of achieving this balance lies in adopting strategies that minimize the harm of downsizing while maximizing the benefits of reform. This begins with a thorough and transparent evaluation of the federal workforce to identify roles and processes that are truly redundant, as opposed to those that are merely inconvenient to reformers. In many cases, inefficiencies can be addressed through retraining and redeployment rather than outright elimination of positions. Investing in upskilling programs allows employees to transition into roles aligned with modern governmental needs, particularly those involving data analysis, technology management, and public engagement.

Another critical approach involves fostering partnerships with the private sector to absorb workforce transitions. Strategic collaborations with industries experiencing growth—such as clean energy, healthcare technology, and digital infrastructure—can provide displaced workers with new opportunities that leverage their existing skills while aligning with broader economic trends. These partnerships also underscore a commitment to supporting workers rather than discarding them as collateral damage in the pursuit of efficiency.

Leadership and communication are pivotal in managing this delicate balance. Policymakers must articulate a clear vision for reform that emphasizes its dual goals: improving government performance and protecting the dignity of its workforce. Transparency in decision-making, coupled with active engagement with labor unions, advocacy groups, and affected communities, fosters trust and mitigates resistance. Public dialogue, facilitated through town halls, online platforms, and other participatory mechanisms, ensures that reforms are shaped by those they affect most directly.

Ultimately, the question of balancing jobs and efficiency is not a binary choice but an opportunity to innovate. It challenges the nation to redefine what a modern, effective, and equitable government looks like in the 21st century. By combining ethical considerations with strategic foresight, the path forward can lead to a government that not only works better but also serves as a model for how to reconcile efficiency with compassion.

Supporting Displaced Workers

Supporting displaced workers is an essential component of fostering a just and equitable transition as governmental reforms like those initiated by DOGE take shape. Efficiency-driven governance, while vital for reducing waste and ensuring fiscal responsibility, must not overlook the human cost associated with workforce reductions or structural realignments. For those whose livelihoods are disrupted, a comprehensive framework of support must be implemented to ensure dignity, opportunity, and economic stability during periods of change.

Central to this endeavor is the acknowledgment of the profound psychological and financial toll displacement can inflict on individuals and families. The sudden loss of employment often disrupts not only income streams but also professional identity, social connections, and long-term aspirations. Addressing these impacts requires a multifaceted strategy that incorporates immediate relief, skill enhancement, and pathways to reintegration into the workforce.

In the short term, safety nets must be robust and responsive. Expanded unemployment benefits, coupled with direct financial assistance for essential needs such as housing, healthcare, and childcare, can provide stability during the transition period. However, such measures should not merely be stopgap solutions. Instead, they must be integrated with proactive career services designed to accelerate reemployment. This includes access to job-matching platforms leveraging AI to connect workers with opportunities that align with their skills, experience, and geographical preferences.

Reskilling and upskilling programs play a pivotal role in preparing displaced workers for the demands of an evolving job market. In collaboration with private industry and educational institutions, government initiatives should offer tailored training programs that reflect emerging economic trends. For instance, if a restructuring effort displaces workers from administrative roles, training in areas such as data analytics, cybersecurity, or green technologies can open doors to new and growing sectors. These programs should be made accessible through flexible learning formats, including online courses, evening classes, and modular certifications, ensuring inclusivity for those balancing familial or financial responsibilities.

Equally critical is fostering an ecosystem of entrepreneurship for those inclined to create their own opportunities. By providing seed funding, mentorship programs, and access to business networks, the government can empower individuals to pursue ventures that not only offer personal fulfillment but also contribute to local economic development. Tax incentives for startups and simplified regulatory pathways can further encourage innovation and job creation among displaced workers.

An effective support framework also involves community-level engagement. Local governments and non-profit organizations are often better positioned to address specific needs within their regions. Partnering with these entities ensures that support mechanisms are context-sensitive and impactful. For example, a rural community facing job losses from federal facility closures might prioritize agriculture-based innovations, while an urban setting could focus on technology hubs or creative industries.

The emotional and mental health of displaced workers should not be overlooked. Comprehensive support services, including counseling and peer support groups, can help individuals navigate the uncertainties of career transitions. Such services should be destigmatized and readily available, fostering resilience and a sense of empowerment among affected individuals.

Transparency in the reform process is vital to maintaining public trust. Displaced workers and their communities deserve clear, honest communication about the reasons for workforce reductions, the steps being taken to mitigate impacts, and the opportunities available for their futures. This openness not only alleviates fears but also reinforces the government's commitment to equitable reform.

Ultimately, the measure of successful governance lies not only in its efficiency but in its humanity. Supporting displaced workers is not merely a moral obligation; it is an investment in the long-term health and stability of society. By equipping individuals with the tools, resources, and opportunities to adapt and thrive, DOGE can set a precedent for reform that harmonizes fiscal discipline with social responsibility, crafting a legacy of progress that truly serves all citizens.

Guarding Against Conflicts of Interest

Guarding against conflicts of interest is a fundamental requirement for ensuring the credibility and integrity of any reform initiative, particularly one as ambitious as DOGE. The potential for conflicts to arise within an efficiency-driven governance model is significant, given the scale of partnerships, the integration of private expertise, and the centralization of decision-making authority. Safeguarding against these risks demands a framework of stringent ethical standards, vigilant oversight, and proactive accountability mechanisms.

At the heart of this endeavor is the principle of transparency. A system designed to eliminate inefficiency and promote trust must itself operate openly, subject to scrutiny from both internal and external stakeholders. To achieve this, DOGE should establish clear protocols for the disclosure of financial interests and affiliations for all individuals and organizations involved in its operations. This includes senior leadership, advisory bodies, and contractors. Publicly accessible records of these disclosures can help illuminate potential biases or overlapping interests that might otherwise go unnoticed, thereby fostering public confidence in the reform process.

The selection of Elon Musk and Vivek Ramaswamy as key figures in DOGE underscores the importance of private-sector collaboration in driving government efficiency. However, their dual roles as private entrepreneurs and public reformers highlight the inherent challenges of navigating overlapping responsibilities. To address these concerns, strict guidelines must delineate the boundaries between their public mandates and private ventures. Contracts awarded by DOGE, particularly those involving industries in which these figures hold stakes, must be subject to rigorous vetting through independent review boards. Such boards, composed of non-partisan experts and citizen representatives, can assess potential conflicts impartially and recommend alternative strategies where necessary.

Oversight extends beyond individual figures to the broader network of private entities engaged in implementing reforms. While partnerships with industry leaders can accelerate innovation, they also risk creating dependency or favoritism that undermines the reform agenda. To counteract this, competitive bidding processes should govern all contract awards, ensuring that decisions are based solely on merit and cost-effectiveness rather than influence or prior relationships. Additionally, ongoing performance evaluations should measure the impact of these partnerships, providing a basis for corrective action if contractual obligations are unmet or if conflicts of interest surface.

The use of artificial intelligence within DOGE's operations further complicates the ethical landscape. AI systems can introduce biases not only through flawed algorithms but also through the influence of the entities that develop and manage these technologies. For example, a private contractor tasked with deploying fraud-detection tools may have a vested interest in prioritizing certain datasets or outcomes that align with their commercial objectives. To mitigate this risk, the development and application of AI within DOGE must adhere to open-source principles where feasible, enabling independent audits of code and decision-making processes. Transparency in this context ensures that AI serves public interest rather than private gain.

Public engagement also plays a crucial role in identifying and addressing conflicts of interest. Empowering citizens to act as watchdogs through platforms that allow for the submission of concerns or evidence of potential conflicts can strengthen accountability. A whistleblower protection program tailored to DOGE's scope should encourage individuals within and outside the organization to report unethical practices without fear of retaliation. This grassroots approach to oversight ensures that ethical breaches cannot be concealed within bureaucratic silos.

Preventing conflicts of interest also requires a cultural shift within the federal apparatus. DOGE must exemplify a commitment to ethics at every level of its hierarchy, embedding principles of fairness, impartiality, and public service into its core identity. Mandatory ethics training for all personnel, from leadership to entry-level staff, can reinforce these values. Such training should not merely outline legal requirements but also delve into the broader moral imperatives of governance, cultivating a workforce attuned to the potential consequences of their actions.

Ultimately, the integrity of DOGE's reforms hinges on its ability to navigate the complex interplay of public and private interests without compromising its mission. A vigilant approach to conflict prevention not only protects against scandals and inefficiencies but also reinforces the public's belief in the transformative potential of government efficiency. By committing to transparency, impartiality, and accountability, DOGE can uphold the democratic ideals it seeks to strengthen and inspire confidence in its vision for a leaner, more responsive government.

Musk, Ramaswamy, and Contractor Relationships

The interplay between Elon Musk, Vivek Ramaswamy, and the contractors engaged under the auspices of DOGE represents both an unprecedented opportunity and a complex challenge in the realm of government reform. Central to the department's mission of fostering efficiency and transparency, these figures bring a potent mix of entrepreneurial expertise, public engagement strategies, and innovative thinking. However, their prominent roles also necessitate a vigilant approach to managing relationships and mitigating the risks of conflicts of interest that could compromise the integrity of the reform process.

Musk's commitment to innovation and technological solutions aligns seamlessly with the goals of DOGE. His advocacy for transparency and accountability has already catalyzed significant advancements in areas such as fraud detection and operational streamlining. Yet, as a prominent business leader with extensive commercial interests in industries ranging from aerospace to artificial intelligence, the potential for perceived or actual conflicts is an unavoidable reality. A similar dynamic applies to Ramaswamy, whose focus on public engagement and participatory governance is complemented by his entrepreneurial background in the pharmaceutical and financial sectors. Together, their leadership epitomizes a modern synergy between private ingenuity and public service. Still, it also underscores the critical need for clear boundaries and oversight mechanisms.

The contractor relationships developed under their stewardship reflect the department's broader reliance on private-sector partnerships to achieve its ambitious objectives. Contractors are often at the forefront of implementing the cutting-edge solutions DOGE advocates, whether through developing machine learning algorithms, enhancing cybersecurity, or optimizing federal workflows. However, these collaborations raise critical questions about favoritism, procurement ethics, and the concentration of influence among a select group of firms.

Establishing trust in these relationships begins with rigorous and impartial procurement processes. Competitive bidding remains a cornerstone of fair contractor selection, ensuring that opportunities are distributed based on merit, capability, and cost-effectiveness. Open calls for proposals, transparent evaluation criteria, and external oversight of award decisions can further fortify this process against undue influence. Such measures not only safeguard the integrity of contract allocations but also reinforce public confidence in DOGE's reform efforts.

Equally important is the ongoing management of contractor performance. Regular audits, progress reviews, and accountability benchmarks ensure that contractors fulfill their obligations without deviation or compromise. These evaluations should be conducted by neutral third parties or interdisciplinary panels with no vested interests in the outcomes. The inclusion of public representatives in such panels could also add a layer of citizen oversight, aligning with DOGE's commitment to participatory governance.

The personal and professional ties of Musk and Ramaswamy to the contractors working with DOGE warrant particular scrutiny. Any overlap between their private enterprises and the department's engagements must be disclosed proactively and subjected to independent review. This is especially relevant when contracts involve technologies or services where these leaders maintain significant market presence. In such instances, recusals from decision-making roles may be necessary to avoid even the perception of bias. Clear guidelines on disclosure and recusal are essential, both to protect the credibility of DOGE and to uphold the ethical standards expected of public leadership.

In navigating these complexities, DOGE must also address the broader implications of its contractor relationships for public accountability. As private firms assume increasingly prominent roles in governance, questions arise about the transparency of their operations and their accountability to the citizenry. Contractors must adhere to the same standards of openness and ethics as DOGE itself, with their processes and outputs made accessible for public review where appropriate. By extending the principles of transparency to its partners, DOGE can ensure that its reform efforts remain grounded in the values of openness and inclusivity.

Ultimately, the relationship between Musk, Ramaswamy, and DOGE's contractors exemplifies the delicate balance between harnessing private-sector dynamism and preserving the impartiality and equity of public service. By implementing robust safeguards, fostering a culture of integrity, and prioritizing transparency, DOGE can navigate these challenges successfully, setting a new standard for collaboration that advances the goals of efficiency, accountability, and public trust. In doing so, it not only strengthens its own operations but also redefines the relationship between government and the private sector in the pursuit of common good.

Ensuring Unbiased Reform

Ensuring that reform efforts are free from bias is not only a moral imperative but also a practical necessity for the success and legitimacy of any governance initiative. Within the ambitious scope of DOGE, the risk of partiality—whether in the application of policies, the allocation of resources, or the design of technologies—can erode public trust and compromise the very principles of fairness and accountability that underlie its mission. Safeguarding against bias requires a proactive, multi-pronged approach rooted in transparency, rigorous oversight, and a steadfast commitment to inclusivity.

The deployment of artificial intelligence as a cornerstone of DOGE's reform strategy underscores the importance of this issue. AI systems, while offering unparalleled opportunities for efficiency, are inherently shaped by the data and assumptions that inform their algorithms. Bias can seep into these systems through historical inequities reflected in training data, poorly defined objectives, or the unconscious preferences of developers. To counteract this, DOGE must prioritize the adoption of inclusive datasets that represent the full diversity of the American populace. Furthermore, algorithms should undergo routine audits by independent third parties to identify and correct disparities in outcomes. These audits must be publicly documented, reinforcing the principle of transparency and providing a framework for continuous improvement.

Human oversight remains a critical component in mitigating algorithmic bias. While automation can streamline decision-making processes, it is imperative that no significant action is taken without the contextual understanding and ethical discernment that only human judgment can provide. DOGE should establish cross-disciplinary review panels to evaluate AI-driven recommendations, ensuring that they align with constitutional values and the broader goals of equity and justice.

These panels, composed of technologists, ethicists, and representatives from diverse communities, can offer a balanced perspective on the implications of automated decisions.

Beyond technology, the human element in policy execution also demands careful scrutiny. Bias in governance often manifests through systemic practices rather than overt actions, making it challenging to identify and address. A comprehensive training program for DOGE personnel, emphasizing unconscious bias awareness and ethical decision-making, can help foster a culture of fairness and impartiality. This training should be iterative, incorporating lessons from past missteps and evolving to address new challenges as they arise.

Public participation is another critical safeguard against bias. DOGE's commitment to democratizing efficiency can be realized through mechanisms that empower citizens to contribute to and critique reform efforts. Platforms for crowdsourcing ideas, public consultation sessions, and accessible channels for grievance redressal ensure that diverse perspectives inform every stage of the reform process. This participatory model not only enhances inclusivity but also acts as a check against the concentration of influence within a narrow demographic or interest group.

To institutionalize impartiality, DOGE must also address structural factors that predispose governance to favoritism. This includes standardizing metrics for evaluating the success and impact of reforms, ensuring that all programs and initiatives are assessed by the same rigorous criteria. These metrics should be developed in consultation with a wide array of stakeholders, encompassing academic experts, industry leaders, and community advocates, to reflect a comprehensive understanding of societal needs and priorities.

Accountability mechanisms play a pivotal role in enforcing unbiased practices. Whistleblower protections should be robust, enabling individuals to report instances of bias or unethical behavior without fear of reprisal. Independent oversight bodies, empowered to investigate and resolve complaints, can further reinforce the integrity of DOGE's operations. These bodies should operate with full autonomy, insulated from political pressures or conflicts of interest, and their findings should be published in a manner accessible to the general public.

Ultimately, unbiased reform is not merely about eliminating disparities; it is about actively cultivating a system that upholds the dignity and rights of every individual it serves. DOGE's success depends on its ability to transcend entrenched inequalities and redefine governance as a force for collective advancement. By embedding impartiality into its foundational practices, the department can set a powerful precedent for reform efforts worldwide, demonstrating that efficiency and equity are not mutually exclusive but are, in fact, mutually reinforcing pillars of good governance.

Ethical AI Deployment

The deployment of artificial intelligence within governance represents one of the most transformative opportunities of the modern era. However, with such transformative power comes profound responsibility. Ethical considerations must form the bedrock of AI integration to ensure that its potential is harnessed not merely for efficiency but for justice, equity, and the protection of democratic principles. Within the framework of DOGE's mission, the ethical deployment of AI is not a peripheral concern but a central tenet that shapes the department's approach to innovation and reform.

The potential for AI to revolutionize government operations lies in its ability to process vast quantities of data, identify inefficiencies, and provide actionable insights with unprecedented speed and accuracy. Yet, the very attributes that make AI so powerful also make it susceptible to misuse or unintended consequences. Algorithms are not neutral; they are the product of human design, imbued with the biases and assumptions of their creators. As such, deploying AI ethically begins with transparency at every stage of its lifecycle, from development to implementation.

Transparency demands that the methodologies and decision-making processes of AI systems be accessible and understandable. This entails open-source development where feasible, enabling independent audits and scrutiny by experts and the public alike. For proprietary systems, stringent guidelines must be established to ensure that the opacity of trade secrets does not obstruct accountability. Furthermore, clear explanations of AI-driven decisions must be provided, particularly in cases where those decisions directly impact citizens' lives. This practice, often referred to as "explainable AI," fosters trust and allows for the redressal of errors or grievances.

Bias within AI systems represents one of the most significant ethical challenges. Whether arising from skewed training datasets or implicit assumptions in algorithmic design, bias can perpetuate and even exacerbate existing inequalities. To mitigate this risk, DOGE must commit to the use of diverse and representative data that reflects the multifaceted realities of the American population. Additionally, algorithms should be regularly tested for disparate impacts across demographic groups, and corrective measures must be taken when biases are identified. These evaluations should not be one-time events but ongoing processes, integrated into the operational lifecycle of every AI system.

The automation of government functions also raises critical questions about the balance of power and the role of human oversight. While AI can streamline processes, it must not supplant the essential human elements of empathy, ethical judgment, and accountability. Decisions with significant societal implications—such as resource allocation, workforce reductions, or policy recommendations—must always be subject to human review. To this end, DOGE should establish interdisciplinary oversight committees, composed of technologists, ethicists, legal experts, and community representatives, to evaluate the ethical implications of AI applications within governance.

Privacy concerns are another cornerstone of ethical AI deployment. The data that fuels AI systems often includes sensitive information about individuals, raising the stakes for ensuring robust safeguards against misuse or breaches. DOGE must enforce strict data privacy protocols, including anonymization techniques, secure storage practices, and clear limitations on data usage. Citizens must be fully informed about how their data is being utilized and afforded the opportunity to opt out where appropriate. This approach not only protects individual rights but also reinforces public confidence in AI as a tool for the public good.

The integration of AI into government processes also necessitates consideration of its impact on the workforce. Automation can displace jobs, creating economic and social disruptions that disproportionately affect certain sectors or communities. Ethical deployment requires that these disruptions be anticipated and addressed proactively. This includes investing in reskilling and upskilling programs, fostering pathways for displaced workers to transition into emerging fields, and ensuring that automation is implemented in a way that complements rather than replaces human labor.

Ultimately, the ethical deployment of AI is not a static goal but an evolving commitment. As technology advances, new challenges and opportunities will arise, demanding vigilance and adaptability. DOGE's approach to AI must therefore be rooted in principles that are enduring yet flexible, capable of guiding decision-making in an ever-changing landscape. By prioritizing transparency, fairness, privacy, and accountability, DOGE can set a global standard for how AI is deployed in service of the public. This standard not only ensures the integrity of the department's reforms but also reaffirms the democratic values that define the nation it serves.

Avoiding Algorithmic Bias

Algorithmic bias represents one of the most pressing challenges in the ethical integration of artificial intelligence within government systems. Left unchecked, biases in algorithms can perpetuate and amplify societal inequalities, undermining trust and fairness in the very systems designed to enhance efficiency and equity. The deployment of AI within DOGE must therefore be guided by robust safeguards to ensure that these tools operate impartially, uphold democratic values, and genuinely serve all citizens.

Bias in algorithms often stems from the datasets used to train them. Historical data, while comprehensive, may reflect systemic inequalities or skewed patterns that influence AI behavior. For instance, employment datasets may reflect historical disparities in hiring practices, leading to algorithms that disproportionately disadvantage certain demographics. Addressing this issue begins with critically evaluating the sources and composition of training data. By curating datasets that are representative of the population and scrutinizing them for embedded biases, DOGE can lay the groundwork for fairer algorithms.

However, the challenge does not end with data selection. Algorithms themselves must be rigorously tested for unintended outcomes. Preemptive audits should simulate a variety of scenarios to identify disparities in the algorithm's recommendations or decisions. For example, an AI system designed to allocate resources must be assessed for how it distributes those resources across different regions, income levels, or demographic groups. These tests should be iterative, adapting as new data and insights emerge to ensure that the system remains equitable over time.

Transparency in the algorithmic decision-making process is equally critical. Understanding how an AI system arrives at its conclusions allows for accountability and correction where needed. Explainability—ensuring that AI systems can provide clear, human-understandable reasons for their decisions—should be a fundamental design criterion for any system deployed by DOGE. This practice not only builds trust but also empowers stakeholders to engage meaningfully with the technology.

Collaboration with diverse stakeholders is another essential element in avoiding algorithmic bias. The development and deployment of AI systems should involve input from technologists, ethicists, policymakers, and representatives from affected communities. This multidisciplinary approach ensures that a broad spectrum of perspectives is considered, minimizing the risk of blind spots in the system's design and implementation. Engaging communities early in the process also fosters a sense of ownership and trust, as citizens see their values and concerns reflected in the outcomes.

Ongoing monitoring and adaptation are vital to addressing biases that may emerge after deployment. Even the most carefully designed algorithms can exhibit unanticipated behaviors when exposed to real-world conditions. Establishing mechanisms for continuous oversight, such as independent review boards and automated monitoring systems, enables DOGE to identify and rectify issues promptly. These measures ensure that fairness is not a one-time consideration but an enduring commitment.

Accountability is central to the success of these efforts. When bias is identified, there must be clear protocols for redress. Whether this involves recalibrating an algorithm, compensating affected parties, or revising policies, the response should be transparent, timely, and proportionate to the impact. Furthermore, individuals impacted by AI-driven decisions should have access to appeals processes, ensuring that they are not left voiceless in the face of technology.

Ultimately, avoiding algorithmic bias is not merely a technical challenge but a reflection of DOGE's broader ethical mandate. By committing to fairness, transparency, and accountability, the department can set a powerful example of how technology can be harnessed to promote justice rather than entrench inequality. In doing so, DOGE reaffirms its role as a steward of progress that respects and elevates the values of democracy and equality at every turn.

Transparency in Automation

Transparency in automation is not merely a technical requirement but a cornerstone of ethical governance, ensuring that systems intended to optimize and streamline public administration remain accountable to the people they serve. In an era where algorithms increasingly influence critical decisions, the imperative for openness in their design and operation has never been greater. DOGE, as a pioneer of government reform through advanced technologies, must embrace a culture of transparency that not only enhances trust but also serves as a model for other institutions.

Automation within public systems carries the promise of improved efficiency, consistency, and scalability. Yet these advantages are accompanied by risks that arise from the opaque nature of many algorithmic processes. Decisions made by automated systems—whether related to resource allocation, eligibility for services, or policy enforcement—often appear inscrutable to those affected. This opacity can breed distrust, particularly when errors or perceived injustices emerge. Transparency acts as a remedy, demystifying how decisions are made and enabling scrutiny that holds systems accountable.

Building transparent automation begins with the design process. Algorithms must be created with clarity of purpose and function, avoiding unnecessary complexity that could obscure their operation. While sophisticated techniques such as machine learning often require intricate computations, their application within public governance must prioritize interpretability. Developers and policymakers should collaborate to ensure that every system can be explained in plain terms, allowing stakeholders to understand not only what decisions are made but why they are made.

Open documentation is a critical element of this transparency. Comprehensive records detailing the development, testing, and deployment of automated systems should be accessible to both internal auditors and the public. These records should include descriptions of the datasets used, the assumptions underlying the algorithmic models, and the intended outcomes. Such documentation not only enhances accountability but also provides a foundation for addressing concerns or criticisms that may arise.

Transparency also requires mechanisms for real-time insight into the functioning of automated systems. Public-facing dashboards or interfaces that provide summaries of system operations, such as the criteria used for decision-making or the patterns observed in outcomes, can demystify the process for citizens. For instance, if an automated system is used to identify instances of waste in federal spending, a transparent summary of flagged transactions, categorized by criteria and regions, could reassure stakeholders that the system operates fairly and effectively.

Engagement with external oversight further strengthens transparency. Independent audits of automated systems— conducted by experts unaffiliated with the developers or operators—offer an impartial assessment of their fairness, accuracy, and alignment with stated objectives. Such audits should be conducted regularly and their findings made public, reinforcing the system's credibility.

Transparency must extend to error management as well. No system, regardless of sophistication, is immune to mistakes or unforeseen consequences. A transparent approach acknowledges this reality and establishes protocols for identifying, disclosing, and rectifying errors. Citizens affected by automated decisions should have clear pathways to appeal or seek redress, with assurances that their grievances will be addressed equitably. Moreover, feedback from these processes should inform continuous improvements to the system, creating a virtuous cycle of refinement.

Public participation in the development and oversight of automated systems is another vital component. Citizens should be invited to contribute their perspectives, particularly when systems are deployed in areas that directly affect their lives. Whether through town halls, advisory panels, or online consultation platforms, such engagement ensures that automation aligns with the values and priorities of the communities it serves.

Ultimately, transparency in automation is about more than compliance or technical precision—it is a reflection of democratic principles in action. By opening the black box of algorithmic governance, DOGE can foster a culture of trust, collaboration, and continuous improvement that exemplifies the very ideals of efficiency and accountability it seeks to advance. In doing so, it not only elevates its own reform efforts but also sets a benchmark for institutions worldwide, demonstrating that technology and transparency can coexist in the pursuit of a more equitable future.

Chapter 6: Technological Foundations

The AI Arsenal

The arsenal of artificial intelligence deployed by DOGE represents a transformative leap in governance, blending technological precision with the vision of a more efficient and accountable federal system. Each tool in this sophisticated collection plays a distinct role, yet together they embody a unified strategy aimed at addressing long-standing inefficiencies, enhancing transparency, and fostering data-driven decision-making.

At the heart of this technological array lies predictive analytics, a cornerstone of DOGE's mission to anticipate challenges before they escalate into crises. By leveraging historical data and advanced modeling techniques, these systems offer insights into potential areas of waste or inefficiency. For instance, predictive algorithms can analyze patterns in federal expenditures, flagging anomalies that might indicate overspending or mismanagement. This proactive approach not only curtails unnecessary costs but also redefines the way government anticipates and resolves operational bottlenecks.

Complementing predictive tools are machine learning algorithms tailored to identify fraud and abuse across federal programs. These systems operate with remarkable agility, sifting through vast datasets to detect subtle irregularities that might elude human oversight. For example, an algorithm designed to monitor procurement processes might highlight discrepancies in contractor billing practices, enabling swift corrective action. Such capabilities underscore AI's potential to safeguard public resources, ensuring that taxpayer dollars are utilized responsibly and transparently.

Optimization algorithms form another critical component of DOGE's AI suite, streamlining processes that have historically been burdened by redundancy and inefficiency. Whether applied to workforce management, logistics, or policy implementation, these tools identify the most effective allocation of resources to achieve desired outcomes. In practice, this might mean recalibrating staff assignments across agencies to maximize productivity or restructuring supply chains to reduce costs and delays. The result is a government that operates with precision and purpose, free from the inertia of outdated practices.

Equally transformative is the integration of natural language processing technologies, which revolutionize the way government interacts with citizens and processes information. Virtual assistants and chatbots equipped with these capabilities can handle routine inquiries, provide real-time updates on services, and even assist in filing forms or resolving disputes. Meanwhile, text analysis tools delve into public feedback, extracting insights from surveys, social media, and other platforms to inform policy decisions. This dynamic interaction fosters a more responsive and citizen-centric governance model.

Data visualization tools further enhance the accessibility and impact of DOGE's AI arsenal. By translating complex datasets into intuitive visual formats, these systems enable policymakers, stakeholders, and the public to grasp insights at a glance. Dashboards that illustrate budget allocations, performance metrics, or the outcomes of reform initiatives empower all parties to engage meaningfully with the data, promoting accountability and trust.

Underlying these advanced tools is a robust infrastructure designed to ensure their reliability, security, and ethical deployment. Cloud-based architectures facilitate seamless data integration across agencies, while encryption protocols and cybersecurity measures protect sensitive information from breaches. Moreover, governance frameworks guide the responsible use of AI, embedding principles of fairness, transparency, and accountability into every aspect of system design and operation.

As this arsenal continues to evolve, the potential for innovation remains boundless. Emerging technologies such as quantum computing, augmented intelligence, and decentralized data platforms hold promise for further enhancing the capabilities and reach of DOGE's AI systems. However, the adoption of these advancements must be tempered by careful deliberation, ensuring that progress aligns with the ethical and constitutional values that define DOGE's mission.

In harnessing this arsenal, DOGE reimagines what government can achieve in the digital age. By combining cutting-edge technology with a steadfast commitment to public service, it not only addresses the inefficiencies of today but also lays the groundwork for a future of informed, equitable, and agile governance.

Tools and Systems for Reform

The transformation of governance envisioned by DOGE relies heavily on a suite of advanced tools and systems designed to implement reforms at unprecedented scale and precision. These technologies, collectively referred to as the operational arsenal, represent the cutting edge of artificial intelligence, data management, and process optimization. Together, they form the backbone of a modernized government that is agile, transparent, and responsive to the needs of its citizens.

At the core of this arsenal are tools designed to collect, analyze, and interpret vast amounts of data. Centralized data integration platforms serve as the foundation, consolidating disparate datasets from across federal agencies into a unified framework. This allows for a holistic view of government operations, enabling analysts and policymakers to identify inefficiencies, track spending, and evaluate program outcomes with unparalleled accuracy. The integration process employs sophisticated algorithms capable of reconciling inconsistencies in data formats, ensuring seamless interoperability across systems.

Machine learning systems play a pivotal role in the analytical dimension of these reforms. These systems are engineered to detect patterns, anomalies, and trends that would otherwise go unnoticed. For example, predictive algorithms assess procurement cycles to forecast potential cost overruns, while fraud detection models monitor transactions for irregularities indicative of waste or corruption. These applications extend beyond mere identification, offering actionable insights that guide corrective measures, ensuring that every dollar spent is justified and effective.

Another indispensable component of this toolkit is the suite of natural language processing technologies. These tools revolutionize how government interacts with citizens and processes unstructured data. Automated document review systems accelerate the evaluation of contracts, regulations, and public feedback, extracting key information and highlighting areas that require attention. Similarly, virtual assistants powered by conversational AI enhance accessibility by providing real-time responses to public inquiries, streamlining services such as tax filing, benefits registration, and regulatory compliance.

Optimization tools further extend the capabilities of DOGE's systems. These algorithms are designed to tackle complex logistical and resource allocation challenges, ensuring that government resources are utilized to their fullest potential. In practical terms, this might involve redesigning supply chains for emergency response efforts, balancing workloads across administrative staff, or optimizing infrastructure maintenance schedules to reduce downtime and costs.

Visualization technologies complement these analytical tools by transforming raw data into intuitive graphical representations. Dashboards equipped with dynamic charts, heatmaps, and predictive timelines empower decision-makers to grasp complex information at a glance. Such interfaces not only facilitate informed policymaking but also promote transparency by making data accessible to the public. Citizens can monitor the progress of reforms, track spending, and evaluate the impact of government initiatives in real time, fostering a culture of accountability.

To ensure the seamless operation of these tools, a robust infrastructure underpins their deployment. Cloud computing platforms provide the scalability necessary to handle the immense data processing demands, while edge computing ensures rapid analysis and response at localized levels. Advanced encryption and cybersecurity measures safeguard sensitive information, maintaining the integrity and trustworthiness of government systems.

While these tools and systems represent remarkable technological advancements, their success depends on the human expertise that guides their deployment. Training programs for government employees focus on building proficiency in these technologies, ensuring that the workforce is equipped to harness their full potential. Moreover, interdisciplinary collaboration among technologists, policymakers, and community leaders ensures that the design and application of these systems align with ethical standards and societal needs.

By integrating these tools and systems, DOGE not only addresses the inefficiencies of the present but also sets the stage for a government that is adaptive, resilient, and innovative. This operational arsenal embodies the promise of technology to not only streamline governance but to elevate it, ensuring that it remains a force for public good in an ever-evolving world.

Machine Learning in Decision-Making

Machine learning has emerged as a transformative force in decision-making processes, reshaping how organizations—particularly government entities—approach problem-solving and policy implementation. Within the framework of DOGE, machine learning is not merely a tool for automation; it is a strategic enabler that brings precision, adaptability, and scalability to governance. By leveraging the unique capabilities of these systems, DOGE has the potential to revolutionize how decisions are made, ensuring they are data-driven, efficient, and equitable.

At its core, machine learning operates by identifying patterns within vast datasets and using these insights to make predictions, classifications, or optimizations. Unlike traditional algorithms, which follow explicitly programmed instructions, machine learning systems learn from data, continuously refining their models to improve performance. This adaptability makes them particularly well-suited for the dynamic and often complex challenges faced by federal agencies.

One of the most significant applications of machine learning in governance lies in predictive analytics. These systems can forecast trends and outcomes with remarkable accuracy, empowering decision-makers to act proactively rather than reactively. For example, in managing public health crises, machine learning models can analyze epidemiological data to predict the spread of diseases, enabling targeted interventions that minimize impact. Similarly, in disaster response, predictive systems can assess risk factors such as weather patterns and infrastructure vulnerabilities, ensuring resources are allocated where they are most needed.

In the realm of fraud detection, machine learning has proven to be a game-changer. By analyzing transactional data, these systems can identify anomalies indicative of fraudulent activity, such as irregular spending patterns or inconsistent reporting. Unlike static rule-based systems, machine learning models evolve to detect new and increasingly sophisticated fraud schemes, safeguarding public funds with a level of vigilance that would be impossible through manual oversight alone.

Another critical application is in optimizing resource allocation. Machine learning algorithms excel at balancing competing priorities and constraints, making them invaluable for budget planning, workforce management, and supply chain logistics. For instance, in the context of federal budgeting, machine learning can simulate the impact of different funding scenarios, identifying strategies that maximize societal benefits while adhering to fiscal constraints. This capability ensures that resources are directed toward programs with the greatest potential for positive impact.

Natural language processing (NLP), a specialized subset of machine learning, has also become an integral part of decision-making processes. NLP systems analyze and interpret human language, enabling governments to extract actionable insights from unstructured data such as public comments, social media posts, and legislative texts. By processing this information at scale, NLP tools can identify emerging public concerns, assess sentiment trends, and even recommend policy adjustments that reflect the priorities of citizens.

Despite these advantages, the deployment of machine learning in decision-making is not without challenges. The quality of outcomes is heavily dependent on the quality of the data used for training these systems. Biases inherent in the data can lead to biased decisions, perpetuating or even exacerbating existing inequalities. Ensuring fairness requires rigorous scrutiny of training datasets, as well as ongoing audits of model performance to identify and mitigate disparities.

Transparency is equally critical. Machine learning models often operate as "black boxes," producing results without providing clear explanations of how those results were derived. This opacity can undermine trust and accountability, particularly when decisions have significant implications for individuals or communities. To address this, DOGE must prioritize the development and deployment of explainable AI systems—models that not only make decisions but also provide interpretable justifications for those decisions. This approach fosters confidence among stakeholders and ensures that machine learning enhances rather than detracts from democratic values.

Another consideration is the ethical use of machine learning. Decisions driven by these systems must align with constitutional principles and societal norms, ensuring that technological progress does not come at the expense of human rights or dignity. Establishing robust governance frameworks that outline acceptable use cases, accountability measures, and redress mechanisms is essential for maintaining the integrity of machine learning applications.

The integration of machine learning into decision-making processes represents a profound shift in how governments can operate, offering tools that are not only more efficient but also more adaptive and insightful than traditional approaches. However, realizing this potential requires a careful balance between innovation and responsibility. By embracing these technologies while adhering to strict ethical standards, DOGE has the opportunity to set a new precedent for governance—one that combines the power of machine learning with the enduring values of transparency, fairness, and public accountability.

Challenges in AI Implementation

The implementation of artificial intelligence within the framework of government reform offers transformative potential, but it also brings with it a set of formidable challenges. These obstacles are not simply technical in nature; they span ethical, operational, and societal dimensions. Addressing these challenges is imperative for DOGE to achieve its vision of a streamlined and accountable federal government while upholding the principles of fairness, transparency, and public trust.

Data privacy and security constitute one of the most pressing concerns. The effectiveness of AI systems is predicated on access to vast datasets, many of which contain sensitive personal information. Without robust safeguards, the risk of breaches, misuse, or unauthorized access increases significantly. Ensuring the integrity of this data requires the implementation of state-of-the-art encryption protocols, secure storage solutions, and stringent access controls. Additionally, data usage policies must be clearly defined and rigorously enforced to prevent exploitation, such as the inappropriate monetization of citizens' information.

Bias within AI systems is another critical challenge. Machine learning models derive their insights from historical data, which often reflects existing inequalities and systemic biases. When these biases are embedded in the training datasets, they risk perpetuating or even amplifying disparities in outcomes. Addressing this requires a multi-layered approach, including the careful curation of training data, the use of fairness-focused algorithms, and ongoing audits to identify and mitigate biases. Moreover, diverse teams of developers and analysts can bring a broader range of perspectives to the design and evaluation process, further reducing the risk of unintentional prejudice.

Transparency in AI decision-making remains an ongoing hurdle. Many AI systems operate as "black boxes," producing outcomes without providing clear explanations of how those outcomes were reached. This opacity can undermine accountability, particularly when decisions have significant impacts on individuals or communities. To counteract this, DOGE must prioritize the development of explainable AI—systems that offer clear, interpretable justifications for their decisions. This not only builds public confidence but also empowers stakeholders to challenge and improve the decision-making process when necessary.

Technological integration across government agencies poses logistical and operational challenges. Many federal systems operate on outdated or incompatible infrastructures, creating barriers to the seamless deployment of advanced AI tools. Modernizing these systems requires substantial investment in both technology and training. Furthermore, the integration process must be meticulously planned to minimize disruptions and ensure continuity of services during the transition.

Resistance to change, both within government institutions and among the public, is another obstacle. The introduction of AI often evokes concerns about job displacement, ethical implications, and the potential overreach of automated systems. Engaging stakeholders through transparent communication and participatory decision-making processes is crucial to addressing these fears. Educational initiatives can demystify AI technologies, highlighting their benefits while addressing legitimate concerns about their implementation.

Resource constraints also play a significant role in limiting the scope and speed of AI adoption. Developing, deploying, and maintaining sophisticated AI systems requires significant financial and human capital. This challenge is compounded by the need to balance investments in AI with other pressing governmental priorities. Strategic partnerships with private sector entities, academic institutions, and non-profits can alleviate some of these resource limitations, fostering innovation while sharing costs and expertise.

Ethical considerations further complicate the implementation of AI in governance. Decisions about where and how to deploy these systems must align with constitutional values and societal norms. Establishing comprehensive ethical guidelines, informed by experts across multiple disciplines, ensures that AI applications uphold human rights, equity, and justice. Mechanisms for oversight and redress are equally important, providing a framework for addressing instances where AI systems fail to meet these standards.

Finally, the challenge of keeping pace with rapid technological advancements cannot be overstated. AI is a constantly evolving field, and systems implemented today may become obsolete or insufficient within a few years. To remain at the forefront of innovation, DOGE must cultivate a culture of continuous learning and adaptability. This includes investing in research and development, fostering collaborations with cutting-edge technology providers, and regularly updating its systems to incorporate new capabilities and insights.

Overcoming these challenges is no small task, but it is essential for realizing the transformative potential of AI in government. By addressing these obstacles head-on, DOGE can set a global standard for ethical, effective, and equitable AI implementation, demonstrating that technology and governance can coalesce to create a future that serves all citizens with integrity and efficiency.

Data Privacy Concerns

Data privacy concerns have emerged as a critical challenge in the integration of artificial intelligence into government systems. The reliance on vast datasets to power machine learning algorithms, optimize processes, and inform decision-making introduces significant risks related to the security and ethical use of sensitive information. For DOGE, which seeks to lead a transformative reform in governance, addressing these concerns is not only a technical necessity but also a fundamental aspect of maintaining public trust and upholding democratic values.

The scope of data privacy issues encompasses both the collection and storage of information as well as its application in AI-driven systems. Government databases often contain highly sensitive personal information, including financial records, healthcare data, and social service histories. The aggregation of such data into centralized systems, while essential for operational efficiency, creates a prime target for cyberattacks. The consequences of a breach are not merely technical but deeply personal, potentially exposing individuals to financial fraud, identity theft, or unwarranted surveillance.

Safeguarding this data begins with robust cybersecurity measures. Encryption technologies must be employed at every stage of data handling, ensuring that information remains secure during storage, transmission, and processing. Multi-factor authentication, access controls, and rigorous monitoring systems are equally crucial to prevent unauthorized access. These measures, however, represent only the first line of defense. A comprehensive data security strategy must also include regular audits, penetration testing, and incident response protocols that enable swift action in the event of a breach.

Transparency is an essential element of addressing privacy concerns. Citizens have a right to understand how their data is collected, stored, and utilized. Clear, accessible communication regarding data usage policies not only fosters trust but also aligns with constitutional principles of accountability and individual rights. Initiatives such as user-friendly privacy dashboards can empower individuals to review and manage their data, granting them greater control over its application.

The ethical use of data presents another layer of complexity. AI systems often analyze vast amounts of personal information to draw inferences, predict behaviors, or allocate resources. While these capabilities enhance the efficiency of governance, they also risk encroaching on privacy and autonomy. For instance, predictive policing models that rely on demographic and location data have faced criticism for perpetuating biases and disproportionately targeting vulnerable communities. To mitigate such risks, DOGE must adopt clear ethical guidelines governing the use of data, ensuring that AI applications respect individual dignity and societal values.

Data minimization principles play a vital role in reducing privacy risks. By collecting only the information necessary for specific purposes and securely disposing of it once those purposes are fulfilled, the government can limit exposure to breaches and misuse. Furthermore, anonymization techniques—such as data masking and differential privacy—can obscure individual identities within datasets while preserving their utility for analysis. These practices strike a balance between harnessing data's potential and safeguarding individual rights.

Collaboration with external stakeholders is critical in navigating the complexities of data privacy. Partnerships with cybersecurity firms, academic researchers, and civil society organizations provide valuable expertise and perspectives, enhancing the government's capacity to address emerging threats. Additionally, engaging with the public through forums and consultations ensures that privacy policies reflect the concerns and expectations of the citizens they aim to protect.

The integration of privacy-by-design principles into AI development and deployment processes further reinforces the commitment to ethical data practices. From the outset, systems should be engineered with privacy as a core consideration, incorporating features such as automated data audits, granular access controls, and built-in anonymization capabilities. These measures not only enhance security but also streamline compliance with regulatory frameworks.

International cooperation is another dimension of data privacy in an interconnected world. As data flows increasingly transcend national borders, aligning with global standards and participating in cross-border agreements becomes essential. DOGE must navigate this landscape carefully, ensuring that its practices meet international norms while safeguarding national interests.

Ultimately, addressing data privacy concerns requires an ongoing commitment to vigilance, adaptability, and innovation. As technologies evolve and new threats emerge, DOGE must remain proactive, continuously refining its policies and practices to uphold the principles of transparency, fairness, and trust. By prioritizing data privacy, the department not only protects individual rights but also strengthens the foundation of its reform agenda, demonstrating that efficiency and ethics are not mutually exclusive but mutually reinforcing pillars of governance.

Technological Hurdles

The integration of artificial intelligence into government systems has the potential to transform governance, but the road to implementation is fraught with technological hurdles that must be addressed to ensure success. These challenges are both structural and systemic, encompassing issues of infrastructure compatibility, scalability, and the evolving nature of AI technologies. Overcoming these obstacles is critical for DOGE to achieve its ambitious vision of a more efficient and accountable government.

One of the most significant barriers lies in the outdated infrastructure of many federal agencies. Legacy systems, designed decades ago, lack the computational capacity and interoperability required for modern AI applications. These systems often operate in silos, preventing seamless data sharing and integration across departments. Addressing this issue requires substantial investment in IT modernization, including the adoption of cloud computing platforms and the development of standardized data exchange protocols. This foundational step not only facilitates the deployment of AI tools but also enhances overall operational efficiency.

Scalability presents another challenge. The implementation of AI systems must accommodate the vast and diverse scope of federal operations, ranging from social services to defense. Ensuring that AI solutions can handle the scale and complexity of these tasks requires robust algorithms, high-performance computing resources, and advanced data management strategies. Pilot programs can serve as a valuable testing ground, allowing for iterative development and refinement before full-scale deployment.

The rapid pace of technological advancement further complicates AI implementation. Tools and algorithms that are state-of-the-art today may become obsolete within a few years, necessitating a commitment to continuous innovation. Staying ahead of this curve requires a proactive approach to research and development, as well as partnerships with leading technology firms and academic institutions. These collaborations can provide access to cutting-edge expertise and resources, enabling DOGE to adapt its systems to emerging trends and challenges.

Another hurdle is the complexity of integrating AI into existing workflows. Government processes are often governed by strict regulations and protocols, which can make the adoption of new technologies cumbersome. Ensuring that AI applications align with these requirements without compromising their functionality demands careful planning and customization. Change management strategies, including comprehensive training programs and stakeholder engagement, are essential to smooth this transition.

Data quality and availability also pose significant challenges. AI systems rely on large, accurate datasets to function effectively, but gaps in data collection and inconsistencies in data formatting can hinder performance. Establishing rigorous data governance frameworks, including protocols for data cleaning, validation, and standardization, is crucial to address these issues. Additionally, fostering a culture of data literacy within government agencies can enhance the quality of data inputs and the effectiveness of AI systems.

Cybersecurity concerns add another layer of complexity. The deployment of AI systems increases the attack surface for cyber threats, necessitating advanced security measures to protect sensitive information and critical infrastructure. This includes not only technical defenses such as encryption and intrusion detection systems but also the implementation of robust policies and practices for incident response and risk management.

Lastly, the deployment of AI technologies must account for ethical considerations and public perceptions. Concerns about surveillance, bias, and automation-induced job displacement can create resistance to AI adoption. Transparent communication about the goals, benefits, and safeguards of AI initiatives is essential to build public trust and support. Ethical frameworks, developed in consultation with diverse stakeholders, ensure that AI systems operate within the bounds of societal values and legal standards.

Overcoming these technological hurdles requires a coordinated effort that combines innovation with pragmatism. By addressing infrastructure limitations, fostering scalability, staying ahead of technological advancements, and prioritizing cybersecurity and ethics, DOGE can navigate the complexities of AI implementation. In doing so, it not only lays the groundwork for a more efficient government but also sets a precedent for how technology can be harnessed to serve the public good.

The Future of AI in Governance

The future of artificial intelligence in governance is poised to redefine the very foundations of how governments operate and engage with their citizens. As AI continues to evolve, its potential applications extend beyond optimization and efficiency into realms that promise greater inclusivity, adaptability, and foresight. For DOGE, this represents both an opportunity and a responsibility to shape a model of governance that balances technological advancement with the enduring values of democracy and equity.

One of the most promising aspects of AI's future in governance lies in its ability to facilitate more dynamic policymaking. Traditional government processes often rely on static models that struggle to account for rapidly changing social and economic conditions. AI's capacity for real-time data analysis enables governments to identify emerging trends, predict outcomes, and adjust policies with unprecedented agility. This could transform areas such as economic planning, where machine learning models might simulate the impact of tax reforms across various demographics, allowing policymakers to optimize for both growth and equity.

Another transformative application is in personalized citizen services. AI-driven systems have the potential to deliver tailored solutions based on individual needs, ensuring that government programs are both effective and accessible. For example, predictive models could identify individuals at risk of falling into poverty and proactively offer support through targeted benefits or job training programs. Similarly, AI-powered virtual assistants could provide real-time guidance on navigating bureaucratic processes, reducing barriers and improving public satisfaction.

The integration of AI into public safety and crisis management is also set to advance significantly. Predictive analytics could be used to anticipate natural disasters, allocate emergency resources more effectively, and even simulate evacuation scenarios to optimize safety measures. In law enforcement, AI could enhance capabilities in areas such as cybercrime detection and evidence analysis, while maintaining strict safeguards to prevent misuse or overreach.

Ethical considerations will play a pivotal role in guiding the future of AI in governance. As these technologies become more deeply embedded in decision-making, the need for transparency, accountability, and fairness will only grow. Governments must prioritize the development of explainable AI systems that provide clear, interpretable insights into how decisions are made. Public trust hinges on the ability to ensure that AI operates without bias, respects privacy, and adheres to the highest standards of ethical conduct.

The democratization of AI is another frontier with transformative potential. By leveraging open-source technologies and fostering public participation, governments can empower citizens to co-create solutions and contribute to decision-making processes. Platforms that allow for crowdsourced feedback, paired with AI's analytical capabilities, could revolutionize how governments understand and respond to public needs. This participatory model not only enhances transparency but also strengthens the social contract between governments and their citizens.

Global collaboration will be essential as AI continues to reshape governance. Shared challenges such as climate change, cybersecurity, and economic inequality demand cooperative approaches that transcend national borders. AI's ability to analyze and synthesize vast amounts of data can facilitate international partnerships, enabling coordinated responses to global crises. However, this also requires the establishment of international standards and frameworks to ensure ethical and equitable use of AI technologies.

Innovation in AI governance is not without its challenges. Issues such as algorithmic bias, data privacy, and the potential for technology-driven inequality must be addressed proactively. Investments in education and workforce development are critical to ensure that the benefits of AI are broadly shared. Equipping individuals with the skills to thrive in an AI-driven world ensures that technological progress translates into societal advancement.

Looking ahead, the role of AI in governance will be defined by its ability to enhance the fundamental principles of democracy, transparency, and accountability. By embracing innovation while upholding ethical standards, DOGE can set a precedent for how governments worldwide can leverage AI to build a more responsive and resilient public sector. This vision of the future is not merely about efficiency but about creating systems that reflect the values and aspirations of the societies they serve, paving the way for a governance model that is both forward-thinking and deeply human-centered.

Innovations on the Horizon

The horizon of innovation in artificial intelligence for governance holds extraordinary possibilities, presenting opportunities to redefine the relationship between government and citizens while addressing some of society's most pressing challenges. These emerging advancements are poised to expand the functionality and impact of AI systems, offering transformative solutions that go beyond the boundaries of current applications. For DOGE, this wave of innovation represents the chance to further its mission of efficiency, transparency, and accountability in unprecedented ways.

One of the most exciting developments on the horizon is the integration of quantum computing into AI systems. Quantum computing, with its unparalleled ability to process complex computations at speeds far beyond traditional systems, could revolutionize predictive analytics in governance. By enabling real-time analysis of massive datasets, quantum-enhanced AI could provide governments with more accurate forecasts for economic planning, climate change mitigation, and resource allocation. This level of insight could allow policymakers to address issues proactively, minimizing risks and optimizing outcomes.

Advancements in ethical AI frameworks are another area of focus. Future systems are being designed to incorporate built-in ethical decision-making protocols, ensuring that AI-driven policies align with societal values and legal standards. These frameworks use interdisciplinary inputs from ethics, law, and sociology to create AI systems that make decisions transparently and equitably, reducing the risk of bias and fostering greater public trust.

The emergence of decentralized AI systems presents an opportunity to democratize access to governance tools. These systems operate on blockchain technology, allowing for secure, distributed decision-making processes that are transparent and resistant to tampering. Decentralized AI could be particularly impactful in participatory governance, enabling citizens to contribute to policy development or budgetary decisions through secure and verifiable platforms. This innovation could foster a more inclusive and engaged society, strengthening the social contract between government and its constituents.

Another significant innovation is the development of multimodal AI systems. These systems combine multiple forms of data input— such as text, images, audio, and video—to provide a comprehensive understanding of complex situations. In governance, multimodal AI could enhance public safety measures by analyzing video feeds for real-time threat detection while simultaneously processing emergency communication logs. Similarly, it could improve urban planning by integrating satellite imagery with demographic and traffic data to design smarter, more efficient cities.

AI-driven simulations and digital twins are also set to play a crucial role in future governance. Digital twins create virtual replicas of physical systems, allowing governments to simulate scenarios and test policy impacts before implementation. For instance, a digital twin of a city could be used to model the effects of infrastructure changes, helping planners optimize designs for sustainability and efficiency. These tools offer a risk-free environment for experimentation, enabling data-driven decisions that minimize unintended consequences.

The next generation of natural language processing promises even deeper integration into citizen interactions. Conversational AI systems, powered by advanced NLP, will provide more nuanced and context-aware communication, making government services more accessible and user-friendly. For instance, virtual assistants could offer multilingual support, ensuring that language barriers do not prevent citizens from accessing essential services. These systems could also be tailored to provide personalized policy updates, keeping citizens informed about initiatives that directly affect them.

The integration of AI into environmental governance is another area ripe for innovation. AI-driven sensors and monitoring systems can track environmental changes with unprecedented precision, providing real-time data on air quality, water resources, and biodiversity. These systems can inform policies aimed at combating climate change and preserving natural ecosystems, ensuring that governance aligns with global sustainability goals.

Finally, the advent of collaborative AI systems, designed to work alongside human decision-makers, represents a paradigm shift in governance. These systems leverage the strengths of both human intuition and machine precision, offering recommendations that are data-driven yet adaptable to context. Collaborative AI can assist policymakers in navigating complex trade-offs, ensuring that decisions are both informed and aligned with public values.

The innovations on the horizon are not merely technical advancements; they are transformative tools that have the potential to elevate governance to new heights. By embracing these developments, DOGE can continue to lead the way in redefining government efficiency, transparency, and engagement, setting a global standard for how technology can serve the public good. In doing so, it affirms the belief that progress and principles can coexist, forging a future where governance is not only more effective but also more equitable and inclusive.

DOGE as a Model for Other Nations

The Department of Government Efficiency represents a groundbreaking paradigm that has the potential to inspire nations worldwide. As governments across the globe grapple with issues of inefficiency, fiscal responsibility, and the need for technological integration, DOGE stands as a testament to the transformative power of innovation combined with accountability. Its model, built on principles of transparency, ethical governance, and participatory reform, offers a blueprint adaptable to diverse political and cultural contexts.

At the core of DOGE's success is its commitment to leveraging artificial intelligence to optimize government operations. By utilizing predictive analytics, machine learning, and natural language processing, DOGE has demonstrated the capacity of AI to reduce waste, enhance decision-making, and improve service delivery. For other nations, particularly those with resource constraints, these technologies offer scalable solutions that can address inefficiencies without requiring extensive infrastructural overhauls. The ability to tailor AI systems to local needs ensures that the lessons of DOGE can be applied in settings ranging from emerging democracies to established bureaucracies.

Another hallmark of the DOGE model is its emphasis on ethical AI deployment. By prioritizing fairness, transparency, and accountability, the department has set a standard for addressing concerns related to bias, privacy, and surveillance. This ethical framework is crucial for fostering public trust, especially in regions where government actions have historically been met with skepticism. Countries seeking to adopt similar reforms must consider embedding such safeguards to ensure that technology serves as a tool for empowerment rather than control.

Participatory governance is another key element of DOGE's appeal as a global model. Through initiatives such as crowdsourced policy development and public accountability dashboards, DOGE has demonstrated the potential of citizen engagement in shaping government priorities. This approach not only strengthens the legitimacy of reforms but also taps into the collective intelligence of the populace, yielding solutions that are both innovative and inclusive. For nations with vibrant civil societies, these mechanisms offer a pathway to deepen democratic practices and enhance governmental responsiveness.

DOGE's structural reorganization of the federal apparatus provides further insights for international adoption. By streamlining agencies, eliminating redundancies, and implementing data-driven oversight, the department has reduced operational bloat while maintaining effectiveness. This aspect of the model is particularly relevant for governments burdened by legacy systems and overlapping jurisdictions. Tailored approaches to restructuring, informed by DOGE's experience, can help other nations achieve similar efficiencies while respecting their unique administrative traditions.

The economic implications of DOGE's model are also worth noting. By cutting unnecessary expenditures and reallocating resources toward high-impact areas, the department has demonstrated a sustainable approach to fiscal management. Nations facing debt crises or economic instability can draw inspiration from DOGE's strategic budgeting practices, which balance cost-saving measures with investments in critical services. Furthermore, the department's use of AI for long-term fiscal planning offers a replicable tool for governments seeking to future-proof their economies against uncertainty.

As a model for other nations, DOGE also underscores the importance of leadership and vision. The collaboration of individuals like Elon Musk and Vivek Ramaswamy highlights the value of bringing entrepreneurial expertise and public engagement to the realm of governance. For countries exploring similar reforms, cultivating partnerships between government, private industry, and civil society can provide the innovative energy needed to drive transformative change.

However, adapting the DOGE model to other contexts requires careful consideration of cultural, political, and institutional factors. The success of such initiatives hinges on tailoring reforms to local conditions, ensuring that they align with the values and priorities of the target population. Furthermore, the international dissemination of DOGE's practices should be accompanied by knowledge-sharing initiatives, capacity-building programs, and collaborative research to support effective implementation.

Ultimately, the global relevance of DOGE lies in its demonstration that governance can be both efficient and principled. By embracing technological innovation while maintaining a steadfast commitment to democratic ideals, DOGE has charted a path that nations worldwide can follow. As governments strive to meet the challenges of the twenty-first century, this model offers not just a roadmap for reform but a vision of what governance can achieve when guided by a genuine commitment to serving the public good.

Chapter 7: Lessons from History

Past Reform Efforts

The journey of reform in governance is paved with historical attempts, each shaped by the unique challenges and opportunities of its time. Past efforts offer valuable insights into the principles and pitfalls of restructuring government systems, providing a foundation for initiatives like DOGE to build upon and refine. Examining these precedents reveals the enduring tension between ambition and practicality, the importance of leadership, and the critical role of adaptability in achieving lasting change.

One of the most notable historical examples is the Grace Commission, established in 1982 under President Ronald Reagan. This ambitious initiative sought to identify inefficiencies in federal operations and recommend cost-saving measures. Over the course of two years, the commission produced a detailed report highlighting areas of waste, redundancy, and potential reform. Among its most significant contributions was the identification of billions of dollars in potential savings through streamlined operations and improved management practices. However, despite its comprehensive analysis, the implementation of its recommendations faced significant political and bureaucratic resistance. Many of the proposed changes, though theoretically sound, were hindered by entrenched interests, insufficient political will, and the complexities of navigating a divided Congress.

Another instructive case is the Government Performance and Results Act (GPRA) of 1993. This legislation marked a shift toward performance-based management in federal agencies, emphasizing the need for clear objectives, measurable outcomes, and accountability. The GPRA succeeded in fostering a culture of strategic planning and evaluation within many agencies, laying the groundwork for subsequent reforms. However, its impact was uneven, with varying levels of adoption and effectiveness across departments. The challenges of aligning goals with budgets and translating strategic plans into actionable improvements highlighted the need for sustained leadership and consistent oversight.

Internationally, the reforms implemented in New Zealand during the 1980s and 1990s stand out as a model of transformative change. Faced with a fiscal crisis and an inefficient public sector, the New Zealand government undertook a comprehensive overhaul of its administrative systems. By embracing market-driven principles, introducing accrual-based accounting, and decentralizing decision-making, it achieved significant cost reductions and improved service delivery. These changes were supported by a strong political consensus and a commitment to transparency, underscoring the importance of aligning reform efforts with a clear vision and robust stakeholder engagement.

The lessons of these past efforts underscore several critical themes for initiatives like DOGE. First, the success of any reform effort depends on the alignment of political will, public support, and institutional capacity. Without this trifecta, even the most well-intentioned plans risk stagnation or reversal. Second, the implementation of reforms must account for the human element—recognizing that resistance, inertia, and competing interests are inherent challenges that require thoughtful strategies to address. Clear communication, inclusive decision-making, and mechanisms for feedback and adjustment are essential to navigating these dynamics.

Technology has emerged as a pivotal enabler of modern reform efforts, offering tools to address many of the limitations faced by earlier initiatives. The use of artificial intelligence, data analytics, and digital platforms has the potential to overcome barriers to transparency, accountability, and efficiency that have long plagued government systems. However, the integration of these technologies must be approached with caution, ensuring that ethical considerations, data privacy, and equity are not compromised in the pursuit of progress.

The experiences of past reform efforts also highlight the importance of scalability and sustainability. Many initiatives falter not in their conception but in their execution, unable to maintain momentum or adapt to changing circumstances. Building mechanisms for continuous improvement, such as iterative evaluation and stakeholder collaboration, ensures that reforms remain relevant and impactful over time.

By drawing on these lessons, DOGE can position itself as a model of reform that not only addresses the inefficiencies of the present but also anticipates the challenges of the future. Its success depends on learning from the triumphs and failures of its predecessors, forging a path that balances ambition with pragmatism, and innovation with accountability. In doing so, it has the opportunity to redefine what is possible in governance and inspire similar transformations around the world.

The Grace Commission and Its Shortfalls

The Grace Commission, formally known as the President's Private Sector Survey on Cost Control, was launched in 1982 under the administration of President Ronald Reagan. Its mission was bold and clear: to identify inefficiencies, reduce waste, and uncover cost-saving opportunities within the sprawling federal government. Comprising business leaders, management experts, and private-sector consultants, the commission represented an unprecedented attempt to apply corporate efficiency principles to public administration. While its findings highlighted critical areas for reform, the initiative ultimately fell short of its transformative aspirations, offering lessons that remain relevant for contemporary efforts like DOGE.

The commission's initial promise was rooted in its comprehensive scope and innovative approach. Tasked with scrutinizing virtually every aspect of federal operations, it generated over 2,400 recommendations, projecting potential savings of $424 billion over three years. These proposals ranged from eliminating redundant programs and consolidating overlapping agencies to modernizing procurement practices and privatizing certain government functions. The scale of its ambition underscored the systemic nature of inefficiency in government and the urgent need for structural reform.

Despite its groundbreaking analysis, the Grace Commission encountered significant obstacles in translating its recommendations into actionable policy. One of the primary challenges was political resistance. Many of the proposed cuts and changes faced opposition from entrenched interests, including bureaucratic entities protective of their mandates and legislators wary of the political repercussions of downsizing or restructuring. Furthermore, the decentralized nature of the federal government complicated efforts to implement uniform changes across diverse agencies with distinct cultures and priorities.

Public skepticism also played a role in curtailing the commission's impact. Critics questioned the applicability of private-sector strategies to government operations, arguing that the goals and constraints of public administration differ fundamentally from those of business. While private enterprises prioritize profitability and efficiency, government must balance these objectives with equity, accountability, and public service delivery. This tension highlighted the need for reforms to be not only economically sound but also politically and socially sustainable.

The commission's reliance on voluntary compliance further limited its effectiveness. Unlike legislative mandates, its recommendations lacked the authority to compel action, leaving their implementation at the discretion of individual agencies and policymakers. While some agencies embraced the proposals, others resisted or ignored them, resulting in uneven progress and missed opportunities for broader impact.

The lessons of the Grace Commission are both cautionary and instructive. Its ambitious scope and private-sector insights underscore the potential for innovative thinking to identify and address inefficiencies. However, its limited success in achieving systemic change highlights the importance of securing political buy-in, fostering public trust, and ensuring enforceability in reform efforts.

For DOGE, these lessons underscore the critical importance of integrating stakeholder engagement into its initiatives. Building coalitions among legislators, agency leaders, and the public can mitigate resistance and ensure that reforms are not only technically feasible but also politically viable. Additionally, pairing visionary recommendations with actionable mandates can bridge the gap between analysis and implementation, ensuring that insights lead to tangible outcomes.

The Grace Commission's legacy is one of both inspiration and caution. Its achievements in exposing inefficiencies and proposing solutions remain a benchmark for government reform efforts. Yet its shortcomings highlight the complexities of navigating the political, cultural, and institutional realities of governance. By learning from these experiences, DOGE can chart a more effective path, blending the analytical rigor of the Grace Commission with the adaptive strategies needed to overcome the challenges of reform in the modern era.

International Case Studies

Examining international case studies of governmental reform provides valuable insights into strategies that have succeeded and challenges that persist. Across the globe, nations have undertaken bold initiatives to address inefficiency, streamline bureaucracy, and enhance public accountability. These efforts, while varied in context and scope, share common threads that underscore the importance of vision, adaptability, and sustained commitment.

New Zealand's public sector reforms in the 1980s stand out as one of the most comprehensive examples of transformative change. Confronted with economic stagnation and a bloated public sector, the New Zealand government embraced a radical overhaul grounded in market-oriented principles. Key reforms included the adoption of accrual-based accounting, the privatization of state-owned enterprises, and a shift to performance-based management within government agencies. These measures significantly improved fiscal discipline, reduced inefficiencies, and heightened accountability. A critical factor in New Zealand's success was its emphasis on transparency and stakeholder engagement, which fostered trust and buy-in from both public servants and citizens. However, critics noted the social costs of some reforms, particularly in areas where privatization led to reduced access to essential services. This highlights the importance of balancing efficiency with equity, ensuring that the pursuit of fiscal goals does not come at the expense of societal well-being.

Singapore offers another compelling example of effective governance reform. Widely regarded as one of the most efficient governments in the world, Singapore's public administration is characterized by its meritocratic ethos, strategic use of technology, and focus on long-term planning. The government's commitment to talent development within the civil service has been pivotal, attracting top talent through competitive salaries and rigorous training programs. Additionally, Singapore's integration of digital technologies—such as its Smart Nation initiative—has streamlined service delivery and enhanced citizen engagement. By leveraging technology to eliminate redundancy and promote transparency, Singapore has demonstrated how innovation can serve as a catalyst for effective governance. Yet, the city-state's model is not without its critics, who point to its centralized decision-making processes and the potential risks of overreliance on data-driven systems.

Scandinavian countries, including Sweden and Denmark, have also implemented notable reforms that prioritize both efficiency and inclusivity. In Sweden, a long-standing commitment to decentralization has empowered local governments to manage resources more effectively while maintaining national oversight to ensure equity and consistency. This approach has fostered innovation at the municipal level, where public administrators are better positioned to tailor solutions to local needs. Similarly, Denmark's "Flexicurity" model has redefined labor market policies, combining flexibility for employers with robust social protections for workers. This balance has enabled Denmark to adapt to economic fluctuations while minimizing social disruption. The success of these reforms underscores the value of participatory governance and the importance of aligning policy goals with cultural values and institutional capacities.

The United Kingdom's reforms under Prime Minister Margaret Thatcher offer a contrasting perspective, characterized by aggressive privatization and deregulation during the 1980s. These measures were aimed at reducing public expenditure and fostering private-sector innovation. While these reforms achieved significant fiscal savings and revitalized certain industries, they also sparked controversy over their social impact, particularly in terms of increased inequality and diminished access to public services. The UK experience serves as a cautionary tale about the risks of prioritizing economic efficiency over social equity, highlighting the need for a nuanced approach to reform.

The lessons from these international efforts are multifaceted and instructive for initiatives like DOGE. First, the importance of aligning reforms with cultural and institutional contexts cannot be overstated. What works in one nation may require significant adaptation to succeed in another. Second, transparency and public engagement are critical to building trust and ensuring the sustainability of reforms. Finally, a balanced approach that integrates efficiency with equity is essential to achieving meaningful and enduring change.

By studying these case studies, DOGE can draw inspiration from successful strategies while avoiding the pitfalls of past efforts. The global landscape of governance reform demonstrates that while challenges are inevitable, they are not insurmountable. With a thoughtful and adaptive approach, it is possible to build a government that is both effective and equitable, serving as a model for others to follow.

Key Takeaways for DOGE

Key lessons emerge from past efforts and international experiences in reforming government systems, offering a roadmap for the success of DOGE. These insights, gleaned from both successes and failures, illuminate the path toward a more efficient, transparent, and accountable federal government. Understanding these lessons is essential not just for implementing change but for ensuring its longevity and impact.

The first critical takeaway is the necessity of aligning reforms with institutional realities. One of the consistent challenges across past initiatives, from the Grace Commission to international case studies, has been the gap between visionary goals and operational feasibility. Reforms must be tailored to the existing political, cultural, and administrative contexts, ensuring they are realistic and adaptable. This requires not only a granular understanding of the structures being reformed but also the foresight to anticipate obstacles and resistance.

Another key lesson is the indispensable role of leadership and stakeholder engagement. Reforms succeed when they are championed by leaders who possess not just the authority but the ability to inspire and mobilize support. Equally important is the inclusion of diverse stakeholders—legislators, civil servants, private-sector partners, and citizens. Collaborative approaches build trust, foster innovation, and create a sense of shared ownership over the reform process.

Transparency and accountability have consistently proven to be foundational pillars for successful reform. Initiatives that prioritize open communication and measurable outcomes gain greater public trust and are less likely to face opposition. Clear metrics for success, paired with regular updates on progress, ensure that reforms are not only implemented but also maintained. This openness also facilitates the identification and rectification of issues before they become insurmountable challenges.

The integration of technology into governance, as demonstrated by modern initiatives, offers transformative potential but must be approached with care. While AI and data analytics provide tools for efficiency and predictive insights, their deployment must adhere to strict ethical standards to avoid exacerbating inequalities or infringing on privacy. Lessons from countries like Singapore emphasize the importance of balancing innovation with safeguards, ensuring that technology serves as a means to enhance governance rather than complicate it.

Financial sustainability is another critical consideration. Reforms that achieve short-term savings but undermine long-term stability risk creating more problems than they solve. Budgetary strategies must be forward-looking, balancing immediate efficiencies with investments in essential services and infrastructure. The fiscal prudence of New Zealand's reforms, which paired cost-cutting measures with strategic reinvestments, serves as an instructive example of achieving this balance.

Finally, resilience and adaptability are paramount. Governments operate in dynamic environments, and reforms must be designed to evolve in response to changing circumstances. Mechanisms for continuous feedback, iterative improvement, and adaptive governance ensure that reforms remain relevant and effective over time. This requires a culture of innovation and a willingness to learn from both successes and setbacks.

By internalizing these lessons, DOGE can position itself as not just a driver of immediate efficiency but as a model of enduring reform. The task ahead is not simply to redesign the structures of government but to redefine how they operate, ensuring that they reflect the principles of democracy, accountability, and service to the public. In doing so, DOGE has the opportunity to transform governance into a force that not only meets the needs of today but anticipates the challenges of tomorrow.

What Worked, What Didn't

The lessons of history provide a nuanced understanding of what has worked in governance reform and what has failed, revealing patterns that can guide contemporary initiatives such as DOGE. Efforts from the past have shown that innovation, leadership, and adaptability are essential ingredients for success, yet they must be balanced against the challenges of resistance, misaligned priorities, and unintended consequences. By dissecting these successes and shortcomings, a clearer roadmap for impactful reform emerges.

One of the most successful strategies in reform efforts has been the focus on transparency and accountability. Initiatives that provide clear metrics for progress and open lines of communication with stakeholders often gain greater traction and public trust. For example, reforms that include public dashboards to track government spending or performance benchmarks have demonstrated that transparency fosters engagement and diminishes skepticism. However, when transparency is absent or poorly executed, it can erode confidence and lead to perceptions of opacity or inefficiency. Past efforts have shown that merely declaring a commitment to openness is insufficient; it must be accompanied by actionable, visible measures that invite public participation and scrutiny.

Another area of proven success lies in leveraging technology to streamline operations. Governments that have invested in data analytics, machine learning, and digital service platforms have seen significant improvements in efficiency and citizen satisfaction. The integration of artificial intelligence into fraud detection and resource allocation, for instance, has demonstrated measurable reductions in waste and errors. However, technological initiatives have occasionally faltered due to inadequate infrastructure, insufficient training, or resistance from stakeholders who fear automation will displace traditional roles. Ensuring that technological advancements are paired with robust training programs and phased implementation plans is critical to overcoming these obstacles.

Leadership has consistently been a determining factor in the success of reform. Visionary leaders who articulate a clear direction, build coalitions, and persist in the face of resistance have driven some of the most impactful changes. Yet, leadership alone is not enough; reforms must also be institutionalized to outlive the tenure of their champions. Efforts that hinge solely on individual personalities or transient political will often falter once those figures exit the stage. This underscores the need for structural reforms that embed changes into the fabric of governance, ensuring their continuity across administrations.

The role of public engagement cannot be overstated. Reforms that actively involve citizens in the decision-making process tend to enjoy greater legitimacy and durability. Participatory models, such as town halls, citizen advisory committees, or crowdsourced solutions, empower individuals to contribute to shaping policies that affect their lives. Conversely, reforms that neglect to consider public input often encounter backlash or apathy, limiting their reach and effectiveness.

Failures in reform often stem from a lack of alignment between goals and execution. Grandiose plans that do not account for institutional inertia or political realities risk becoming mired in bureaucracy. Historical examples have shown that reforms need to be scalable and adaptable, starting with pilot programs or incremental changes that build momentum and demonstrate value before expanding nationwide. This approach not only minimizes risk but also creates a foundation of success upon which further reforms can be built.

The interplay between efficiency and equity is another critical lesson. While efforts to streamline operations and cut costs are often necessary, they must be balanced against the need to protect vulnerable populations and preserve access to essential services. Reforms that prioritize efficiency at the expense of equity risk alienating key constituencies and exacerbating social inequalities, undermining the very legitimacy they aim to strengthen.

Finally, one of the most common pitfalls in reform efforts is the failure to anticipate and mitigate resistance. Whether from entrenched interests within the bureaucracy, political opponents, or skeptical citizens, opposition can derail even the most well-conceived initiatives. Building coalitions, engaging in transparent dialogue, and demonstrating quick wins to garner support are essential strategies for overcoming such resistance.

By learning from these lessons, DOGE can navigate the complexities of governance reform with greater precision and resilience. Success will depend not only on the boldness of its vision but also on its ability to execute with humility, adaptability, and a commitment to serving the public good.

Building a Legacy

Building a legacy requires more than the successful implementation of reforms; it demands a lasting impact that transcends political cycles and leaves a framework for future governance to emulate. For DOGE, the vision of transforming the federal government into a more efficient and transparent system is not just about addressing the inefficiencies of today but creating a foundation for sustainable innovation and accountability. Achieving this involves a deliberate focus on institutionalizing reforms, fostering adaptability, and ensuring the principles of transparency and efficiency are deeply ingrained in the fabric of governance.

One of the most critical aspects of leaving a legacy is embedding reforms into the institutional framework of the federal government. Temporary measures or changes tied solely to the vision of current leadership risk being dismantled once political priorities shift. To avoid this, reforms must be codified into policy and supported by mechanisms that ensure their longevity. This could include statutory mandates, oversight bodies, and performance benchmarks that make reversing progress difficult without clear justification. By establishing these safeguards, DOGE ensures that its impact remains durable and resistant to the whims of political change.

Adaptability is another cornerstone of a lasting legacy. Governance operates within a dynamic landscape, where technological advancements, societal expectations, and global challenges constantly evolve. Reforms that are rigid or narrowly focused may struggle to remain relevant as circumstances change. DOGE's success lies in creating systems that are not only effective but also flexible—able to incorporate new tools, address emerging issues, and respond to unforeseen crises. Regular assessments and iterative improvements ensure that these systems continue to serve the public interest in an ever-changing environment.

The cultivation of a culture of efficiency and transparency within the government workforce is equally essential. Beyond structural changes, the mindset and values of the individuals who operate within these systems play a pivotal role in sustaining reforms. Comprehensive training programs, clear communication of goals, and incentives for innovation and accountability can foster an environment where efficiency and transparency are not merely objectives but core principles that guide everyday operations. This cultural shift, once achieved, becomes a self-sustaining force for improvement.

Public trust is fundamental to building a legacy, and it must be earned through consistent demonstration of accountability and effectiveness. DOGE's commitment to transparency, such as through public dashboards and citizen engagement platforms, reinforces the legitimacy of its efforts. By keeping citizens informed and involved, the department not only builds trust but also strengthens the social contract between government and the governed. This trust is a critical component of any lasting legacy, as it ensures public support for reforms even as leadership transitions occur.

Leadership transitions pose one of the most significant risks to the continuity of reforms. New administrations often bring different priorities, which can threaten the progress achieved by their predecessors. DOGE's approach to mitigating this risk involves creating bipartisan buy-in and aligning its goals with universal principles of good governance. By framing its reforms as nonpartisan initiatives aimed at improving the efficiency and effectiveness of government for all citizens, DOGE increases the likelihood of their continuation across political divides.

Finally, the legacy of DOGE must extend beyond the immediate impact on the federal government. As a model of innovative governance, it has the potential to inspire similar efforts at the state, local, and even international levels. By documenting its processes, sharing best practices, and engaging with global reform initiatives, DOGE can amplify its influence and contribute to a broader movement toward more efficient and accountable governance worldwide.

A true legacy is measured not only by what it achieves but by the foundation it lays for future progress. For DOGE, this means institutionalizing reforms, fostering a culture of adaptability and transparency, and building the trust and support needed to sustain its impact over time. By prioritizing these elements, DOGE can ensure that its vision of a leaner, more effective government endures, setting a standard for governance that serves as a beacon for generations to come.

Ensuring Reforms Outlive the Administration

Ensuring that reforms outlive the administration that initiates them is one of the most critical challenges in governance. While bold initiatives may succeed in transforming structures and policies in the short term, their long-term viability often depends on the ability to institutionalize change and foster a culture that embraces continuous improvement. For DOGE, the goal of creating a legacy that endures beyond its founding years necessitates strategic planning, stakeholder engagement, and robust mechanisms to safeguard its principles and achievements.

A primary consideration in preserving reforms is the establishment of structural safeguards that embed changes within the institutional framework. By codifying reforms into legislation or formal policy, DOGE can ensure that its initiatives cannot be easily reversed or undermined by subsequent administrations. Statutory mandates provide a level of permanence that transcends political cycles, making it more challenging for opponents to dismantle progress without clear and compelling reasons. Additionally, integrating reforms into the operational guidelines of federal agencies ensures that they become part of the standard operating procedure, further cementing their place within the governmental ecosystem.

The creation of oversight bodies plays a crucial role in maintaining accountability and continuity. Independent entities tasked with monitoring the implementation and outcomes of reforms can provide an objective assessment of their effectiveness. These bodies, insulated from political influence, ensure that reforms remain aligned with their original intent and continue to deliver measurable benefits. Their findings can also serve as a valuable resource for future administrations, offering insights into what works and what requires refinement.

A culture of adaptability within the government workforce is another vital element. Reforms often face resistance from those accustomed to established routines or skeptical of change. Addressing this requires not only comprehensive training programs but also efforts to instill a mindset that values innovation and efficiency. Recognizing and rewarding employees who embrace these principles can foster a culture where reforms are seen not as transient initiatives but as the foundation of a modern, effective government.

Public trust is indispensable for the longevity of reforms. Citizens must perceive these changes as beneficial and equitable to sustain their support over time. Transparency and consistent communication are key to building this trust. By maintaining open channels for public feedback, DOGE can demonstrate its commitment to accountability and responsiveness, reinforcing its legitimacy. Furthermore, visible successes—such as reductions in wasteful spending, streamlined services, or improved responsiveness—serve as tangible proof of the value of reforms, galvanizing public and political support.

The alignment of reforms with bipartisan principles is another critical strategy. Initiatives perceived as partisan are more likely to be discarded by opposing administrations, regardless of their effectiveness. By framing its efforts as nonpartisan solutions to universally recognized challenges, DOGE can foster broader political consensus. Highlighting the alignment of reforms with constitutional principles and shared national values strengthens their appeal across ideological divides.

Building coalitions among stakeholders also contributes to the durability of reforms. Engaging legislators, civil society, and the private sector in the reform process creates a sense of shared ownership and investment in its success. When reforms are supported by a diverse array of interests, they are less vulnerable to reversal, as dismantling them risks alienating key constituencies.

Finally, the establishment of adaptive feedback loops ensures that reforms remain relevant and effective. Regular evaluations, informed by data analytics and stakeholder input, provide opportunities to refine and improve policies. This iterative approach allows reforms to evolve in response to changing circumstances, preserving their utility and impact over time.

By prioritizing these strategies, DOGE can create a framework that not only implements transformative changes but also ensures their endurance. The ultimate measure of success lies not in the immediate outcomes but in the sustained impact of these reforms, shaping a government that remains efficient, accountable, and responsive for generations to come. In achieving this, DOGE sets a precedent for governance that transcends its origins, becoming a model of enduring reform.

Chapter 8: Public and Political Dynamics

Navigating Political Resistance

Navigating political resistance is one of the most complex challenges faced by any reform initiative, particularly those that seek to transform deeply entrenched systems. For DOGE, the stakes are heightened by the ambitious scope of its mandate and the polarized nature of contemporary politics. Overcoming opposition requires not only strategic acumen but also an unwavering commitment to the principles of transparency, inclusivity, and accountability that define its mission.

Political resistance often stems from competing interests and ideological divides. Within any government, the introduction of sweeping reforms disrupts established power dynamics, prompting pushback from stakeholders who perceive a threat to their influence or priorities. Legislators, bureaucrats, and external interest groups may each present unique challenges, from legislative gridlock to bureaucratic inertia and public campaigns against perceived overreach. To navigate these obstacles, it is essential to identify the underlying concerns driving opposition and address them proactively.

Building coalitions across party lines is critical to mitigating resistance and ensuring the sustainability of reforms. While partisan divides may seem insurmountable, focusing on shared goals—such as fiscal responsibility, improved public services, or national security—can create common ground. Engaging legislators and policymakers early in the process, providing clear data to support proposed changes, and emphasizing the bipartisan benefits of reform help to reduce friction and foster collaboration.

Transparency is another powerful tool in countering political opposition. Resistance often thrives in environments of uncertainty, where stakeholders may exaggerate potential risks or question the motives behind reform efforts. By maintaining open communication, sharing detailed plans, and offering regular updates on progress, DOGE can build trust and counter misinformation. Publicly accessible dashboards, policy briefs, and town hall discussions further enhance credibility, demonstrating a genuine commitment to accountability.

Addressing the concerns of bureaucratic stakeholders requires a nuanced approach. Resistance from within the federal workforce often arises from fear of job losses, increased workloads, or diminished autonomy. A comprehensive strategy for workforce engagement can alleviate these fears by emphasizing the value of reform for both employees and the institution as a whole. This includes providing training and support for those affected by changes, offering clear pathways for career advancement, and highlighting success stories where reforms have improved efficiency without compromising job security.

External opposition, such as that from special interest groups or advocacy organizations, presents additional challenges. These groups often leverage public sentiment and media narratives to amplify their objections, making it essential for DOGE to actively shape the public discourse. Strategic media engagement, supported by compelling evidence of the benefits of reform, can counteract negative messaging and build broader support. Furthermore, inviting external stakeholders to participate in advisory roles or consultative processes can reduce adversarial dynamics and foster a more collaborative atmosphere.

The role of public opinion cannot be understated in navigating political resistance. Reforms that enjoy widespread popular support are significantly more resilient to opposition. Mobilizing grassroots campaigns, leveraging social media, and partnering with community organizations are effective ways to galvanize public backing. When citizens feel informed and included in the process, they are more likely to advocate for reforms, exerting pressure on policymakers to align with public sentiment.

Another essential component in addressing political resistance is the ability to demonstrate early wins. Quick, visible successes—such as measurable cost savings, streamlined services, or reduced processing times—can validate the need for reform and build momentum. These achievements serve as tangible proof of the benefits of change, providing leverage to counteract skepticism and opposition.

Ultimately, navigating political resistance requires a balance of persistence, diplomacy, and adaptability. By fostering inclusivity, maintaining transparency, and demonstrating the tangible benefits of its initiatives, DOGE can overcome opposition and lay the groundwork for lasting, transformative change. The journey is as much about building trust and consensus as it is about implementing policy, ensuring that the reforms not only succeed in the present but endure into the future.

Bipartisan Challenges

The challenges of achieving bipartisan cooperation in advancing governmental reforms often stem from the inherent tensions of a polarized political landscape. In pursuing transformative changes, such as those championed by DOGE, navigating the complexities of bipartisan dynamics requires both strategic finesse and a steadfast commitment to the greater good. Achieving meaningful progress necessitates addressing competing priorities, fostering trust, and framing reforms in a manner that transcends partisan divides.

At the heart of bipartisan challenges is the fundamental reality that political parties often prioritize divergent goals and constituencies. For one side, the emphasis may be on reducing government size and spending, while the other may focus on equity and protecting essential services. Reforms that appear to align too closely with one ideology risk alienating the opposing party, creating an environment where collaboration becomes difficult. This underscores the importance of presenting initiatives as nonpartisan solutions to shared challenges, emphasizing their potential to benefit all citizens regardless of political affiliation.

The historical record reveals both successes and failures in overcoming such divides. Initiatives that have succeeded in garnering bipartisan support typically share common traits: a clear articulation of benefits, evidence-based proposals, and an inclusive approach that engages stakeholders across the political spectrum. For example, reforms aimed at reducing redundancy or cutting wasteful spending often appeal to fiscal conservatives, while improving transparency and citizen engagement resonates with advocates of democratic accountability. By identifying and emphasizing these shared values, reformers can build bridges that unite disparate factions around common objectives.

Transparency plays a pivotal role in diffusing partisan tensions. By openly sharing data, methodologies, and anticipated outcomes, DOGE can preempt accusations of bias or hidden agendas. Publicly accessible dashboards, regular updates on progress, and opportunities for stakeholders to contribute to the reform process further enhance credibility. When all parties have access to the same information and a voice in the conversation, the space for mistrust and misinformation diminishes.

Another strategy for overcoming bipartisan challenges involves leveraging the influence of respected mediators. Leaders and organizations with credibility across party lines can serve as advocates for reform, articulating its benefits in ways that resonate with diverse audiences. These mediators can also facilitate dialogue, helping to resolve conflicts and build consensus. Their involvement lends an additional layer of legitimacy to the initiative, making it harder for critics to dismiss or undermine.

Public opinion serves as a powerful force in breaking through partisan resistance. Reforms that enjoy broad public support place pressure on legislators to prioritize the will of their constituents over partisan considerations. Mobilizing grassroots campaigns, conducting opinion polls, and amplifying citizen voices through media and public forums create a groundswell of support that is difficult for policymakers to ignore. Demonstrating that reforms align with the values and aspirations of the electorate further strengthens their appeal.

Bipartisan challenges are not merely obstacles but opportunities for innovation and compromise. They compel reformers to refine their proposals, ensuring they address the concerns of a broad constituency. This iterative process often results in solutions that are not only more palatable to both sides but also more robust and effective. By embracing the tensions inherent in bipartisan dynamics as a constructive force, DOGE can achieve reforms that are not only successful but enduring.

Ultimately, the ability to navigate these challenges depends on maintaining a focus on the ultimate goal: creating a government that is more efficient, accountable, and responsive to the needs of its people. By prioritizing inclusivity, transparency, and evidence-based decision-making, DOGE can transcend partisan divides, demonstrating that meaningful reform is not only possible but essential for the nation's progress.

Building Alliances

Building alliances is an essential component of any meaningful reform effort, particularly one as ambitious as the transformation spearheaded by DOGE. Alliances provide the foundation for consensus, the leverage to overcome resistance, and the resilience to endure political and operational challenges. At its core, alliance-building is about recognizing shared interests, fostering trust, and creating partnerships that transcend divisions, ensuring that reforms are both effective and sustainable.

The process of forming alliances begins with identifying stakeholders who have a vested interest in the success of the initiative. These include policymakers, civil servants, private sector leaders, and citizens. Each group brings unique perspectives and resources that can strengthen the reform effort. For instance, legislators contribute the authority to enact policy changes, while civil servants offer operational expertise. Engaging the private sector introduces innovative solutions and technological capabilities, and the support of citizens ensures legitimacy and momentum.

Establishing trust is fundamental to successful alliance-building. Trust is earned through transparency, inclusivity, and consistent follow-through on commitments. By openly sharing goals, methodologies, and progress, DOGE can demonstrate its credibility and integrity. Creating platforms for dialogue, such as forums, workshops, and advisory councils, ensures that all voices are heard and that stakeholders feel genuinely included in the decision-making process.

Shared goals are the cornerstone of strong alliances. While individual stakeholders may have different priorities, finding common ground creates a unifying purpose. For example, the desire for fiscal responsibility unites diverse groups, from those focused on reducing wasteful spending to those advocating for effective resource allocation. Similarly, the commitment to transparency appeals to both accountability advocates and those seeking to rebuild public trust. Emphasizing these shared objectives allows DOGE to bridge ideological and institutional divides.

Alliances are also strengthened by tangible benefits that align with the interests of stakeholders. Policymakers, for example, are more likely to support reforms that deliver measurable outcomes they can present to constituents. Civil servants are more inclined to embrace changes that enhance efficiency without jeopardizing their roles. The private sector sees value in reforms that create a more predictable regulatory environment, while citizens prioritize reforms that lead to better services and lower costs. By clearly articulating how reforms address these priorities, DOGE can secure broader and deeper support.

Collaboration with external organizations and experts further bolsters alliance-building efforts. Partnerships with think tanks, academic institutions, and non-governmental organizations bring additional credibility and expertise to the reform process. These entities can provide research, policy recommendations, and independent evaluations that enhance the quality and impact of the initiatives. International partnerships offer opportunities to learn from analogous efforts in other nations, adapting best practices to the unique context of U.S. governance.

Flexibility is critical in maintaining alliances over time. As reforms progress, the landscape of interests and priorities may shift. DOGE must remain attuned to these changes, adjusting its approach to ensure continued alignment with stakeholder needs. Regular communication and feedback loops are essential to identifying and addressing emerging concerns, reinforcing the sense of partnership and shared purpose.

Finally, the success of alliances depends on their resilience in the face of challenges. Opposition, setbacks, or unforeseen obstacles are inevitable in any reform effort. Strong alliances provide a support system that sustains momentum during difficult times. By fostering a culture of mutual respect and collaboration, DOGE can ensure that its alliances endure and adapt, remaining a pillar of its transformational agenda.

In building alliances, DOGE lays the groundwork for a reform movement that is inclusive, credible, and resilient. These partnerships amplify its reach, enhance its legitimacy, and ensure that its vision of a more efficient and accountable government is realized. Through collaboration and shared commitment, DOGE not only achieves its goals but also sets a powerful example of how governance can transcend divisions and unite stakeholders in pursuit of the common good.

The Media's Role

The media holds an unparalleled position of influence in shaping the public discourse surrounding government reforms. Its role in informing, scrutinizing, and amplifying policy initiatives cannot be overstated, particularly when addressing a transformative endeavor like DOGE. The interplay between reform efforts and media coverage often determines the success or failure of public engagement and trust. As such, DOGE must navigate this dynamic carefully, leveraging the media as both an ally and a platform for transparency.

The media's ability to shape public perception stems from its reach and authority. News outlets, whether traditional or digital, serve as the primary source of information for much of the population. They interpret complex policies, highlight successes and failures, and provide a forum for debate. For DOGE, which seeks to reimagine the federal government's efficiency, ensuring accurate and balanced reporting is critical to building public understanding and support. Misrepresentation or sensationalism in media coverage can erode trust and amplify opposition, underscoring the importance of proactive communication strategies.

Transparency is the cornerstone of a constructive relationship with the media. By maintaining open lines of communication and providing timely access to information, DOGE can foster goodwill and reduce the likelihood of misinformation. Regular press briefings, detailed reports, and responsive media liaisons demonstrate a commitment to accountability. Additionally, making data and outcomes publicly accessible allows journalists to ground their reporting in evidence, reducing the scope for speculation or bias.

The role of investigative journalism, while occasionally adversarial, is equally important. Journalists who delve into the complexities of reforms often uncover inefficiencies or unintended consequences that might otherwise go unnoticed. For DOGE, embracing this scrutiny as an opportunity for improvement rather than a threat reinforces its credibility. Constructive criticism from the media can serve as a catalyst for refining policies and addressing blind spots, demonstrating a willingness to adapt in pursuit of the public good.

Social media platforms further amplify the media's role, creating new opportunities and challenges for DOGE. While traditional outlets provide depth and context, social media enables real-time dissemination of information to a global audience. However, the speed and scale of social media also facilitate the spread of misinformation and polarizing narratives. To counteract this, DOGE must actively engage with these platforms, providing clear, concise, and accurate messaging. Platforms like Twitter, Facebook, and Instagram can be leveraged to share updates, highlight successes, and respond to concerns, fostering a direct connection with the public.

The media also plays a vital role in contextualizing reforms within broader societal trends. By framing DOGE's initiatives as part of a global movement toward efficient and accountable governance, journalists can help the public appreciate the significance and urgency of these efforts. Comparative analyses, expert interviews, and feature stories humanize the impact of reforms, making them more relatable and compelling.

Partnerships with media organizations can further enhance DOGE's outreach efforts. Collaborating on educational campaigns, documentaries, or special reports provides an opportunity to deepen public understanding of the issues at stake. Such partnerships should prioritize neutrality and objectivity, ensuring that the content remains informative rather than propagandistic.

However, the media's influence is not without challenges. Sensationalist coverage, driven by the need for ratings or clicks, can distort public perceptions. Complex policies may be reduced to oversimplified soundbites, leading to misconceptions. DOGE must counteract this by emphasizing clarity in its communications, preempting misunderstandings with straightforward explanations and accessible visuals.

Ultimately, the media's role is to serve as a bridge between the government and the public, fostering transparency, accountability, and engagement. For DOGE, embracing this relationship with openness and strategic intent can transform the media into a powerful ally. By shaping a narrative of progress, trust, and innovation, the department not only enhances its public image but also strengthens the foundation of its reform agenda, ensuring that its message resonates across diverse audiences and endures in the collective consciousness.

Shaping Public Perception

Shaping public perception is a nuanced and pivotal component of any large-scale reform initiative. For DOGE, the ability to influence and inform public opinion is not merely a supplementary aspect of its mission; it is a fundamental driver of success. Perception shapes reality in the realm of governance, where public trust and engagement can determine whether reforms are embraced, resisted, or rendered inert. In this context, DOGE must carefully craft its narrative, ensuring that its message resonates with diverse audiences and fosters a collective understanding of its goals.

Public perception is inherently influenced by the clarity and accessibility of information. Complex government reforms, while necessary, often fail to capture public support due to their perceived opacity or technical jargon. DOGE's success depends on translating its objectives and achievements into clear, relatable terms that resonate with the everyday experiences of citizens. This involves not only simplifying the language but also framing the reforms within stories that highlight tangible benefits—reduced wait times, streamlined services, or cost savings reinvested in public programs.

The use of storytelling is particularly effective in shaping perception. Narratives that feature real-life examples of how reforms improve individual lives are more compelling than abstract statistics or policy analyses. A family that experiences faster processing for social benefits, a small business owner who navigates reduced red tape, or a community revitalized through better resource allocation are all powerful testaments to the value of DOGE's initiatives. These stories personalize the impact of reforms, transforming them from distant bureaucratic processes into meaningful changes that citizens can see and feel.

Transparency is another cornerstone in cultivating a positive public perception. DOGE must ensure that its operations, successes, and challenges are shared openly with the public. This includes providing regular updates through accessible platforms such as public dashboards, press releases, and community forums. Transparency not only builds trust but also preempts misinformation, reducing the likelihood of opposition fueled by misunderstandings or speculation. When citizens feel informed, they are more likely to support reforms and advocate for their continuation.

Social media has emerged as a critical tool in shaping public perception, offering an unprecedented ability to engage with citizens directly. Platforms such as Twitter, Facebook, and Instagram allow DOGE to disseminate updates in real time, respond to concerns, and foster interactive dialogue. However, the immediacy of social media also presents risks, such as the rapid spread of misinformation. To mitigate this, DOGE must maintain a robust presence, actively countering false narratives with accurate information and clear explanations.

The role of thought leaders and influencers in shaping public perception should not be underestimated. Engaging respected figures from academia, industry, and civil society can amplify DOGE's message, lending it additional credibility and reach. These individuals can act as intermediaries, interpreting complex reforms for their audiences and endorsing the initiative's goals. Similarly, partnerships with community organizations and local leaders provide grassroots channels for spreading awareness and garnering support.

Another critical aspect of shaping perception is addressing skepticism proactively. Reforms of the scale and ambition pursued by DOGE inevitably face questions about feasibility, fairness, and unintended consequences. Anticipating these concerns and providing evidence-based responses not only mitigates resistance but also reinforces the department's commitment to accountability. For example, preemptively showcasing safeguards against waste or bias in AI-driven systems demonstrates a proactive approach to potential challenges.

The media also plays a significant role in influencing public opinion. By fostering positive relationships with journalists and news outlets, DOGE can ensure that its story is told accurately and compellingly. This includes providing journalists with clear, concise materials, facilitating access to data, and offering opportunities for interviews with key figures involved in the reforms. Positive coverage in trusted outlets further legitimizes DOGE's efforts, helping to solidify public support.

Ultimately, shaping public perception is not about manipulation but about engagement and education. By fostering a transparent, inclusive, and relatable narrative, DOGE can ensure that its vision of a more efficient and accountable government is not only understood but embraced. Public perception, when managed thoughtfully, becomes a powerful force for change, aligning citizens and policymakers in a shared commitment to transformative reform.

Addressing Misinformation

Addressing misinformation is an essential challenge for any reform initiative, particularly one as ambitious and transformative as DOGE. The prevalence of false narratives, whether born of misunderstanding, intentional distortion, or political opposition, poses a significant threat to the public trust that underpins successful governance. Misinformation can obscure the true objectives of reforms, sow division, and undermine the very foundations of constructive debate. For DOGE to achieve its mission, it must adopt a proactive, transparent, and strategic approach to countering misinformation, ensuring that the public and stakeholders have access to accurate, credible, and timely information.

The first line of defense against misinformation is a robust and transparent communication strategy. DOGE must establish itself as a trusted source of information by consistently sharing clear and factual updates on its goals, processes, and achievements. This requires the use of accessible language that demystifies complex policy details, avoiding jargon that might alienate or confuse the audience. By presenting data and outcomes in straightforward terms, DOGE empowers citizens to form informed opinions based on verifiable evidence rather than conjecture.

A critical component of combating misinformation is speed. False narratives, particularly in the digital age, can spread rapidly across social media platforms and news outlets. DOGE must be equipped to respond swiftly, addressing inaccuracies before they gain traction. This involves monitoring online discussions and media coverage, identifying emerging misinformation, and deploying corrective measures promptly. Social media platforms, with their vast reach and immediacy, play a dual role as both the source of misinformation and a vehicle for dispelling it. By maintaining an active presence on these platforms, DOGE can engage directly with users, clarifying misconceptions and redirecting the narrative toward truth.

Collaboration with trusted intermediaries enhances the effectiveness of misinformation countermeasures. Partnerships with journalists, fact-checking organizations, and academic institutions provide additional credibility to corrective efforts. Journalists who understand the nuances of DOGE's initiatives can accurately report and contextualize information, while independent fact-checkers can verify claims and address falsehoods with impartial authority. Academic studies and reports further reinforce the legitimacy of DOGE's narrative by grounding it in research and expert analysis.

Public education is a powerful tool for inoculating against misinformation. By fostering media literacy and critical thinking, DOGE can empower citizens to recognize and question dubious claims. Educational campaigns, workshops, and partnerships with schools and community organizations can equip individuals with the skills to discern credible sources from unreliable ones. This proactive approach not only counters specific instances of misinformation but also builds a more informed and resilient citizenry.

Transparency and accessibility extend to addressing the root causes of misinformation. Some false narratives stem from genuine confusion or gaps in understanding. By identifying and addressing these gaps, DOGE can preempt the spread of inaccuracies. This might involve clarifying the rationale behind certain policies, providing detailed explanations of implementation strategies, or addressing common concerns in a public forum. Open dialogue fosters trust and reduces the likelihood of misinterpretation.

Countering intentional misinformation—efforts deliberately designed to mislead—requires a more assertive approach. Legal and regulatory frameworks play a role in holding perpetrators accountable, particularly in cases of coordinated disinformation campaigns. Additionally, DOGE can work with social media platforms to flag and de-prioritize false content, ensuring that accurate information takes precedence in public discourse.

The most effective response to misinformation is a consistent track record of honesty, transparency, and accountability. When the public perceives DOGE as a trustworthy institution, its statements and corrections carry greater weight. Building this reputation involves not only correcting misinformation but also acknowledging and addressing legitimate critiques. This demonstrates a commitment to fairness and continuous improvement, further solidifying public trust.

In the fight against misinformation, DOGE has the opportunity to set a new standard for government communication. By prioritizing transparency, fostering collaboration, and empowering citizens with knowledge, it can not only protect its reforms from distortion but also contribute to a broader culture of informed and engaged democratic participation. Through these efforts, DOGE reinforces the principle that truth is the foundation of effective governance and public trust.

Mobilizing Public Support

Mobilizing public support is the cornerstone of transformative governance reform. For an initiative like DOGE to succeed, public backing must extend beyond passive approval to active engagement. The public's role in shaping, endorsing, and driving reforms ensures that changes reflect collective priorities, withstand political resistance, and endure as a foundational part of governance. Building such support requires a strategic blend of transparency, engagement, and communication that resonates with citizens and inspires their commitment.

Public support begins with understanding. Complex reforms can often feel distant or abstract to citizens, making it essential to bridge the gap between policy and everyday impact. DOGE must articulate its goals and achievements in terms that are relatable and tangible. For instance, demonstrating how streamlined processes lead to faster services or how cost-saving measures free resources for critical programs personalizes the reforms, transforming them from abstract initiatives into visible benefits.

Active engagement is equally critical. By involving citizens in the reform process, DOGE fosters a sense of shared ownership and accountability. Town hall meetings, online forums, and participatory platforms allow individuals to voice concerns, contribute ideas, and see their input reflected in outcomes. This participatory approach not only strengthens trust but also leverages the collective wisdom of diverse perspectives, enhancing the quality and relevance of reforms.

Transparency is paramount in building and maintaining public trust. DOGE must ensure that its operations, challenges, and outcomes are communicated openly and consistently. Regular updates through accessible channels—such as public dashboards, newsletters, and media briefings—demonstrate accountability and invite scrutiny. When citizens feel informed and included, they are more likely to support and advocate for the reforms.

Storytelling is a powerful tool in mobilizing support. Highlighting real-life examples of individuals or communities positively impacted by DOGE's initiatives brings the abstract to life. Whether it's a small business owner navigating simplified regulations or a family benefiting from streamlined social services, these stories humanize the reforms and make their benefits undeniable. Personal narratives resonate more deeply than statistics, fostering emotional connections that drive public advocacy.

Harnessing the power of social media expands the reach and immediacy of engagement efforts. Platforms like Twitter, Facebook, and Instagram provide opportunities to connect with citizens in real time, share updates, and address misconceptions. By creating dynamic, interactive content—such as infographics, videos, and live Q&A sessions—DOGE can make its message more accessible and engaging, especially to younger demographics. Social media also allows for targeted outreach, ensuring that specific communities or interest groups receive tailored information that speaks to their unique concerns.

Partnerships amplify efforts to mobilize support. Collaborating with community organizations, advocacy groups, and local leaders provides credibility and extends reach. These partners often serve as trusted intermediaries, translating complex reforms into relatable messages and rallying their constituencies. Similarly, engaging influential figures from various sectors—academia, business, entertainment—adds authoritative voices to the conversation, bolstering public confidence in DOGE's mission.

Education campaigns play a vital role in fostering long-term support. By equipping citizens with the knowledge to understand and evaluate reforms, DOGE empowers them to become informed advocates. Workshops, public service announcements, and educational materials distributed through schools, libraries, and community centers create a more informed electorate capable of engaging constructively with the reform process.

Addressing skepticism head-on is critical to solidifying support. Reforms often face criticism or doubt, whether from political opponents, interest groups, or wary citizens. DOGE must anticipate these concerns and respond with evidence-based rebuttals, demonstrating the thoughtfulness and integrity of its approach. Acknowledging legitimate critiques and outlining steps to address them further strengthens credibility, showing that the initiative is responsive and adaptive.

Mobilizing public support is not merely a step in the reform process but an ongoing endeavor. As DOGE progresses, maintaining public enthusiasm and involvement ensures that its mission remains relevant and impactful. By fostering transparency, engagement, and trust, DOGE transforms citizens from passive observers into active partners in creating a more efficient and accountable government. This collective effort not only secures the success of current reforms but also lays the groundwork for a more participatory and responsive democracy.

Grassroots Campaigns

Grassroots campaigns are the lifeblood of democratic reform, transforming broad ideals into actionable movements driven by the energy and commitment of everyday citizens. For DOGE, these campaigns represent a vital pathway to mobilizing public support, fostering a sense of collective ownership, and embedding its reforms into the national consciousness. Grounded in the principles of participation and empowerment, grassroots efforts are uniquely positioned to bridge the gap between policy and people, making governance more tangible and accessible.

The essence of grassroots campaigns lies in their capacity to connect with individuals on a deeply personal level. By rooting messages in local contexts and addressing specific community concerns, these campaigns ensure that the broader goals of DOGE resonate across diverse demographics. For example, while national-level discussions may focus on cutting inefficiencies, a grassroots effort might emphasize how these reforms will improve the turnaround time for small business permits or reduce waste in local infrastructure projects. By tailoring messages to the lived experiences of communities, DOGE can transform abstract policies into relatable and urgent calls to action.

Engaging local leaders is a cornerstone of effective grassroots organizing. Trusted figures within communities—whether they are educators, business owners, clergy, or activists—serve as natural conduits for spreading information and rallying support. Their endorsement lends credibility to DOGE's initiatives, while their networks provide avenues for amplifying the message. Equipping these leaders with clear, accessible materials and opportunities to participate in reform discussions ensures they are not only advocates but also informed partners in the movement.

Grassroots campaigns thrive on visibility and interaction. Community events, such as town hall meetings, workshops, and neighborhood forums, provide platforms for dialogue and collaboration. These gatherings allow citizens to voice concerns, ask questions, and contribute ideas, fostering a sense of inclusion and agency. They also serve as opportunities for DOGE to showcase its commitment to transparency and responsiveness, reinforcing trust and enthusiasm for its goals.

Digital tools have revolutionized grassroots organizing, enabling campaigns to reach wider audiences with greater efficiency. Social media platforms, email newsletters, and community apps create opportunities for real-time engagement, while crowdfunding platforms can support localized initiatives tied to larger reforms. These tools also allow for the collection of feedback and data, enabling DOGE to gauge public sentiment, identify emerging issues, and adapt its strategies accordingly. Online petitions, video testimonials, and virtual town halls further extend the reach of grassroots efforts, ensuring that even the most geographically dispersed communities can participate.

Storytelling is a powerful element of grassroots campaigns. Personal narratives that highlight the transformative impact of DOGE's reforms are especially effective in galvanizing support. Whether it's a family that has benefited from faster access to essential services or a local entrepreneur thriving in a streamlined regulatory environment, these stories humanize the reforms and make their significance undeniable. Sharing these stories through media outlets, social platforms, and community networks creates a ripple effect, inspiring others to join the movement.

Sustainability is a critical consideration in grassroots organizing. Campaigns must build momentum that extends beyond initial enthusiasm, evolving into long-term commitments. This requires cultivating a cadre of dedicated volunteers and organizers who can maintain outreach efforts, organize follow-up activities, and ensure continuity. Providing these individuals with training, resources, and recognition reinforces their engagement and strengthens the campaign's foundation.

Grassroots efforts are also uniquely equipped to counter skepticism and misinformation. By fostering direct, person-to-person communication, they create opportunities for genuine dialogue and clarification. Local organizers can address concerns with empathy and context, dispelling doubts and building bridges of understanding that large-scale media campaigns might struggle to achieve.

Ultimately, grassroots campaigns embody the democratic ethos at the heart of DOGE's mission. They empower individuals to take an active role in shaping their government, transforming abstract policy discussions into collective action. By investing in these campaigns, DOGE not only mobilizes support for its immediate goals but also lays the groundwork for a more engaged and participatory citizenry, ensuring that its reforms are both embraced and sustained for generations to come.

Using Social Media Effectively

Effectively leveraging social media is indispensable for DOGE in reaching diverse audiences, shaping narratives, and fostering engagement in a digital era dominated by interconnected platforms. The vast reach of social media provides a unique opportunity to amplify DOGE's message, ensuring that its vision of governance reform resonates widely and inclusively. However, the medium's immediacy and influence demand a carefully crafted strategy to maximize its potential while mitigating its challenges.

The essence of using social media effectively lies in its ability to democratize information. Platforms like Twitter, Facebook, Instagram, and LinkedIn offer direct communication channels between DOGE and the public, bypassing traditional gatekeepers such as mainstream media. This unfiltered access enables DOGE to share updates, address concerns, and highlight achievements in real time, fostering a sense of immediacy and transparency. By creating a steady stream of informative, engaging, and accessible content, DOGE can establish itself as a credible and relatable voice in the public sphere.

Audience segmentation is a critical component of a successful social media strategy. Each platform caters to distinct demographics with unique preferences for consuming information. For example, Twitter's fast-paced environment is ideal for brief updates and direct engagement, while Instagram's visual focus lends itself to infographics, testimonials, and behind-the-scenes glimpses of reforms in action. LinkedIn, with its professional audience, offers a space to discuss policy implications and thought leadership, while platforms like TikTok can engage younger audiences through creative and relatable content. Tailoring the tone, format, and messaging to fit the platform ensures that DOGE's communication resonates with each audience segment.

Engagement is the cornerstone of social media's effectiveness. Unlike traditional media, where information flows in one direction, social media thrives on interaction. DOGE must actively engage with its audience by responding to comments, answering questions, and participating in online conversations. These interactions humanize the organization, demonstrating a commitment to listening and addressing public concerns. By fostering a sense of dialogue rather than monologue, DOGE strengthens trust and encourages active participation.

Visual storytelling is particularly powerful on social media. Images, videos, and graphics have a higher likelihood of capturing attention and conveying messages quickly. Short videos showcasing the impact of reforms, infographics breaking down complex policies, and photo essays highlighting community stories are effective ways to make DOGE's initiatives tangible and compelling. Authenticity is key—content that feels genuine and human resonates more deeply than overly polished or corporate-style messaging.

Transparency and accountability must remain central to DOGE's social media presence. By sharing both successes and challenges openly, the organization builds credibility and counters potential criticism. Regular updates on progress, explanations of decision-making processes, and acknowledgments of areas for improvement demonstrate a commitment to integrity and continuous improvement. This transparency not only enhances trust but also preempts misinformation by ensuring that accurate information is readily available.

The speed and scale of social media also make it an essential tool for addressing misinformation. When false narratives emerge, DOGE can use its platforms to issue clarifications promptly, ensuring that the truth is accessible and amplified. Pinning corrected information to the top of profiles, linking to detailed explanations, and collaborating with trusted third-party fact-checkers enhance the effectiveness of these efforts.

Campaigns that invite participation further amplify DOGE's reach and impact. Hashtag-driven initiatives, challenges, and user-generated content campaigns encourage citizens to share their perspectives and experiences related to governance reform. These campaigns not only extend the visibility of DOGE's message but also create a sense of collective ownership and community around its mission.

Metrics and analytics provide valuable insights into the success of social media efforts. Monitoring engagement rates, reach, sentiment analysis, and audience demographics helps DOGE refine its strategy, ensuring that its content remains relevant and impactful. Continuous adaptation based on data-driven insights ensures that the organization remains agile and effective in a rapidly evolving digital landscape.

Ultimately, using social media effectively is about fostering connection and engagement. By embracing the medium's potential to inform, inspire, and involve, DOGE can build a vibrant and engaged community that champions its vision for governance reform. Social media becomes not just a tool for communication but a platform for collaboration, empowering citizens to join in shaping a government that is transparent, efficient, and truly responsive to the needs of its people.

Chapter 9: Global Implications of DOGE

Setting a Precedent

Setting a precedent is not merely about achieving reform within the confines of a single nation; it is about establishing a model that demonstrates the transformative potential of efficiency-driven governance on a global stage. DOGE represents a convergence of vision, technology, and participatory governance that challenges traditional paradigms, offering a bold template for rethinking the role of government in the twenty-first century. By exemplifying transparency, adaptability, and innovation, DOGE sets a standard that other nations can look to as they confront their own challenges of bureaucratic inefficiency, fiscal responsibility, and public trust.

A defining aspect of setting a precedent is DOGE's commitment to ethical and participatory reform. Unlike historical efforts that focused narrowly on cost-cutting or restructuring without regard for public engagement, DOGE integrates the citizenry into its processes, ensuring that the reforms reflect shared values and priorities. This inclusivity not only strengthens the legitimacy of its initiatives but also provides a powerful counter-narrative to the perception that government operates in isolation from its people. Other nations observing DOGE's approach may see this participatory model as a pathway to addressing their own governance challenges, balancing efficiency with accountability.

The technological underpinnings of DOGE further solidify its status as a trailblazer. By leveraging artificial intelligence and advanced analytics to detect inefficiencies, predict outcomes, and optimize resource allocation, DOGE demonstrates the transformative potential of technology in governance. These innovations allow for real-time decision-making and adaptability, enabling the government to respond swiftly to emerging needs without compromising long-term goals. For nations grappling with outdated systems and resource constraints, DOGE's technological framework offers a replicable model that bridges the gap between ambition and capacity.

Transparency is another pillar of DOGE's precedent-setting influence. By maintaining open communication channels, public dashboards, and accessible reporting mechanisms, DOGE ensures that its operations are visible and accountable. This transparency builds trust among citizens and stakeholders, creating a feedback loop that reinforces public confidence and institutional integrity. For governments worldwide, adopting similar transparency measures could be transformative, fostering trust and mitigating corruption—a pervasive challenge in many political systems.

The economic implications of DOGE's reforms also contribute to its potential as a global model. By identifying and eliminating wasteful spending, reallocating resources to high-impact areas, and implementing data-driven budgeting practices, DOGE demonstrates that fiscal responsibility and effective governance are not mutually exclusive. Nations facing economic instability or unsustainable debt could look to DOGE's example as a blueprint for achieving financial stability without sacrificing essential services or public welfare.

DOGE's emphasis on ethical AI deployment sets an important standard for the integration of technology in public administration. By addressing concerns such as algorithmic bias, data privacy, and equitable access, DOGE ensures that technological innovation enhances rather than undermines democratic principles. This ethical framework is particularly relevant in a global context where the rapid adoption of AI often outpaces the development of safeguards. DOGE's approach serves as a reminder that progress must be tempered with prudence, offering a model for responsible innovation.

As DOGE continues to demonstrate the efficacy of its reforms, it positions itself as a catalyst for international collaboration. Sharing best practices, facilitating knowledge exchanges, and engaging in dialogues with other nations amplify its influence, creating a network of reform-minded governments committed to efficiency and accountability. This global ripple effect not only strengthens the legitimacy of DOGE's efforts but also contributes to a broader movement toward effective governance as a universal aspiration.

In setting a precedent, DOGE redefines what is possible in governance. Its integration of transparency, technology, and participatory engagement creates a blueprint for transformation that extends beyond national boundaries, offering lessons and inspiration to governments worldwide. By establishing itself as a model of innovation and integrity, DOGE not only achieves its mission but also lays the foundation for a new era of governance that is truly of, by, and for the people.

International Interest in DOGE's Model

The global interest in DOGE's model for governance reform underscores its groundbreaking potential as a catalyst for systemic efficiency and accountability. Across nations, leaders and policymakers grapple with their own bureaucratic inefficiencies, unsustainable fiscal policies, and the erosion of public trust. The Department of Government Efficiency, with its bold vision and innovative methods, has sparked curiosity and admiration, positioning itself as a beacon of modern governance that transcends borders.

The allure of DOGE lies in its holistic approach, which integrates cutting-edge technology, participatory governance, and fiscal discipline. Governments worldwide recognize the potential of these strategies to address their unique challenges. The use of artificial intelligence for fraud detection, resource allocation, and predictive budgeting demonstrates how technology can revolutionize administrative processes. This aspect alone has prompted international delegations and think tanks to study DOGE's implementation, exploring how AI-driven tools can be adapted to varying political, cultural, and institutional contexts.

Beyond its technological advancements, DOGE's commitment to transparency and public engagement resonates deeply with nations seeking to rebuild trust between governments and their citizens. In regions where opacity and corruption have long hindered progress, DOGE's emphasis on open communication and accountability offers a compelling alternative. Its public dashboards, citizen feedback mechanisms, and crowdsourced solutions provide a model for bridging the gap between the state and its constituents, fostering a governance style rooted in inclusivity and responsiveness.

Economic imperatives further drive interest in DOGE's model. Countries facing mounting debt or inefficiencies in resource management are particularly drawn to its cost-cutting measures and strategic budgeting practices. By demonstrating how to eliminate redundancies, streamline operations, and reallocate savings to high-impact programs, DOGE presents a roadmap for achieving fiscal stability without sacrificing essential services. This resonates with nations in both developing and developed economies, where balancing growth and responsibility remains a perennial challenge.

The ethical considerations embedded in DOGE's framework also contribute to its international appeal. In an era where concerns about the misuse of technology and the centralization of power dominate global discourse, DOGE's proactive stance on ethical AI deployment and safeguarding data privacy sets a valuable precedent. These principles align with the aspirations of many governments to innovate responsibly, ensuring that technological progress enhances rather than undermines democratic values.

Multilateral organizations and alliances, such as the United Nations and the Organization for Economic Cooperation and Development, have taken note of DOGE's transformative potential. Discussions around integrating aspects of its framework into global governance initiatives underscore its relevance as a template for collective action. Workshops, conferences, and academic studies have begun to dissect the lessons DOGE offers, with a particular focus on its adaptability across different governmental structures and economic landscapes.

However, the international spotlight also brings challenges. Skepticism about the scalability and transferability of DOGE's model remains a topic of debate. Critics question whether the conditions that enabled DOGE's success—such as its leadership, political backing, and access to advanced technologies—can be replicated elsewhere. Addressing these concerns requires a nuanced approach, emphasizing the principles behind DOGE's strategies rather than advocating for a one-size-fits-all replication.

To maximize its global impact, DOGE has initiated partnerships with international bodies and foreign governments interested in adopting similar reforms. These collaborations facilitate knowledge exchange, pilot projects, and capacity-building efforts, ensuring that the principles of efficiency and accountability are tailored to local needs. Such partnerships not only enhance DOGE's credibility but also foster a spirit of shared learning and innovation.

The international interest in DOGE underscores a growing recognition that governance must evolve to meet the demands of a complex, interconnected world. By setting a standard for what is possible, DOGE inspires others to reimagine their approaches to governance, proving that efficiency, transparency, and citizen empowerment are not merely ideals but achievable realities. In doing so, DOGE establishes itself not just as a national success story but as a global exemplar of reform.

Comparative Analysis of Global Efficiency Models

Conducting a comparative analysis of global efficiency models reveals valuable insights into how various nations have approached the challenges of governance reform. While each country operates within its unique cultural, political, and economic context, the underlying principles of transparency, accountability, and innovation often emerge as universal drivers of success. By examining these models, DOGE can identify strategies that resonate with its vision while avoiding pitfalls encountered elsewhere.

New Zealand stands out as a pioneer of public sector reform, particularly for its groundbreaking efforts in the 1980s. Confronted with economic stagnation and mounting public debt, the government implemented sweeping changes aimed at reducing inefficiency and enhancing fiscal responsibility. Key reforms included the adoption of accrual-based accounting, the privatization of state-owned enterprises, and the introduction of performance-based management in public services. These measures not only streamlined operations but also improved transparency, enabling citizens to hold their government accountable. However, critics argue that the social costs of privatization were not adequately mitigated, highlighting the need for reforms to balance efficiency with equity.

Singapore offers another compelling example, often cited for its exemplary public administration. The city-state's emphasis on meritocracy, long-term planning, and the strategic use of technology has resulted in a government widely regarded as efficient and responsive. Initiatives such as the Smart Nation program, which integrates digital technology across all facets of governance, exemplify how innovation can enhance citizen engagement and service delivery. Singapore's approach underscores the importance of investing in talent development, as seen in its rigorous recruitment and training programs for public servants. Yet, its highly centralized governance model raises questions about replicability in larger or more diverse nations.

The Scandinavian countries, particularly Sweden and Denmark, provide a contrast by prioritizing decentralization and inclusivity. Sweden's system of devolved governance empowers local municipalities to manage resources and deliver services, fostering innovation and accountability at the community level. Similarly, Denmark's "flexicurity" model combines labor market flexibility with robust social protections, enabling economic adaptability without sacrificing worker welfare. These approaches demonstrate the value of aligning reforms with cultural norms and institutional capacities, ensuring that changes are both effective and sustainable.

In contrast, the United Kingdom's experience with austerity-driven reforms during the 2010s offers cautionary lessons. While efforts to reduce public expenditure succeeded in cutting deficits, they also led to significant reductions in essential services, prompting widespread public dissatisfaction. The UK's approach illustrates the risks of pursuing efficiency without adequately addressing the social implications, emphasizing the need for reforms to prioritize outcomes that enhance, rather than diminish, public trust.

The United States has its own historical examples, including the Grace Commission of the 1980s, which sought to identify and eliminate waste within the federal government. While the commission highlighted numerous inefficiencies and proposed hundreds of cost-saving measures, many recommendations were not implemented due to political resistance and a lack of sustained momentum. This underscores the importance of aligning reform efforts with political will and ensuring that proposals are actionable within the existing legislative and administrative framework.

By synthesizing these international lessons, DOGE can refine its strategies to maximize impact and sustainability. The importance of clear communication, stakeholder engagement, and adaptability emerges as a common thread across successful models. Transparency and accountability mechanisms not only enhance public trust but also create a foundation for continuous improvement. Additionally, balancing efficiency with equity ensures that reforms are inclusive and garner broad support.

Global comparisons also reveal the transformative potential of technology in governance. From New Zealand's data-driven budgeting systems to Singapore's AI-enhanced service delivery, technology plays a pivotal role in enabling efficiency and responsiveness. For DOGE, leveraging these innovations while adhering to ethical standards will be critical in setting a benchmark for responsible and effective governance.

As DOGE charts its path forward, the global landscape of efficiency models serves as both a guide and a mirror, reflecting the possibilities and challenges of transformative reform. By integrating best practices and learning from the experiences of others, DOGE can not only achieve its ambitious goals but also contribute to a broader movement toward accountable, transparent, and innovative governance worldwide.

Diplomatic and Economic Effects

The diplomatic and economic effects of DOGE's implementation extend far beyond the borders of the United States, reshaping its global standing and influencing international relations. As a pioneering model of efficiency and accountability, DOGE offers both practical outcomes and symbolic resonance, showcasing a renewed commitment to responsible governance. This dual impact strengthens the U.S.'s position as a leader in global innovation and diplomacy, while also recalibrating its economic relationships with allies and trade partners.

On the diplomatic front, DOGE demonstrates the potential of governance reform to address shared challenges in an interconnected world. As nations face similar issues of inefficiency, debt, and citizen dissatisfaction, the successful implementation of DOGE provides a compelling example of proactive problem-solving. This enhances the credibility of the U.S. in international forums, positioning it as a thought leader in modern governance. Countries eager to adopt similar reforms may look to the U.S. for guidance, creating opportunities for collaboration, knowledge exchange, and the export of best practices.

This leadership role fosters stronger alliances, as nations with shared values align themselves with the principles underpinning DOGE. Transparency, fiscal responsibility, and technological innovation resonate deeply in a global context where these attributes are increasingly seen as markers of stability and reliability. By championing these ideals, the U.S. not only reinforces its moral authority but also builds goodwill among partners who value these same principles.

Economically, the ripple effects of DOGE's success are profound. Streamlining government operations and eliminating wasteful spending create a more efficient allocation of resources, which strengthens the U.S. economy and increases its competitiveness on the global stage. A leaner, more agile federal government reduces the fiscal burden on taxpayers, potentially lowering tax rates and stimulating economic growth. These reforms also enhance investor confidence, attracting both domestic and international investments eager to engage with a stable and forward-thinking market.

Trade relationships stand to benefit as well. A more efficient U.S. government can negotiate and implement trade agreements with greater precision, ensuring mutual benefits and reducing bureaucratic delays. This agility enhances the U.S.'s reputation as a reliable partner, fostering stronger economic ties and expanding market opportunities for American businesses abroad. Furthermore, the demonstration of fiscal discipline through DOGE bolsters the U.S. dollar's status as a global reserve currency, reinforcing economic stability and influence.

DOGE's emphasis on technological innovation creates additional economic opportunities. By integrating artificial intelligence and advanced analytics into government operations, the U.S. positions itself as a hub for cutting-edge research and development. This attracts global talent and partnerships, strengthening the nation's technological ecosystem. Additionally, the technologies and methodologies developed under DOGE can be exported to other countries, creating new markets for American innovation and expertise.

However, these diplomatic and economic gains are not without challenges. The visibility of DOGE's success may provoke criticism from adversarial nations wary of U.S. influence, leading to potential tensions in international relations. Additionally, the scalability of DOGE's model across diverse governance systems requires careful adaptation, as not all nations possess the institutional capacity or cultural alignment to replicate its approach seamlessly.

To navigate these complexities, the U.S. must adopt a collaborative stance, offering technical assistance and fostering partnerships that emphasize mutual benefits. This includes engaging with multilateral organizations to disseminate the lessons of DOGE while respecting the sovereignty and unique needs of individual nations. By framing DOGE as a shared endeavor rather than an imposition, the U.S. can mitigate resistance and build a coalition of reform-minded governments.

The broader economic and diplomatic effects of DOGE reflect the interconnected nature of modern governance. As the U.S. redefines its internal operations, the external ramifications ripple outward, influencing global perceptions and relationships. Through careful stewardship of these effects, DOGE not only transforms the federal government but also strengthens the U.S.'s role as a leader in a world increasingly defined by collaboration, innovation, and shared responsibility.

Strengthening the U.S.'s Global Position

Strengthening the United States' global position through the implementation of DOGE reflects a broader ambition: demonstrating that governance reform is not only a domestic imperative but also a pillar of international leadership. By pioneering innovative strategies to enhance efficiency, accountability, and fiscal responsibility, the U.S. reasserts its role as a standard-bearer for effective governance in an increasingly interconnected world. This renewed strength resonates diplomatically, economically, and strategically, enhancing the nation's influence and fostering a renewed sense of purpose on the global stage.

At its core, DOGE's success underscores the capacity of the United States to adapt and lead in addressing universal challenges. Inefficiency, bureaucratic stagnation, and fiscal irresponsibility are not confined to any one nation; they are global phenomena that undermine public trust and economic stability. By tackling these issues head-on, the U.S. sets a compelling example of proactive governance, inspiring other nations to embark on similar journeys. This leadership not only bolsters America's credibility but also positions it as a mentor and ally to governments seeking to modernize their own administrative structures.

In the realm of diplomacy, DOGE serves as a powerful narrative of reinvention. For decades, perceptions of the U.S. as a global leader have been shaped as much by its governance model as by its military and economic might. The establishment of DOGE and its measurable achievements reaffirm the nation's commitment to innovation and accountability. This is particularly significant in an era when global leadership is increasingly judged by soft power—its ability to inspire, influence, and collaborate on shared challenges. Through initiatives such as knowledge-sharing workshops and international partnerships, the U.S. can leverage DOGE to deepen alliances and foster cooperative reform efforts.

Economically, DOGE's influence strengthens the U.S.'s standing as a hub of stability and opportunity. The reduction of wasteful spending, combined with streamlined government operations, generates confidence among international investors and trade partners. This economic revitalization bolsters the U.S. dollar, reinforcing its status as the global reserve currency and ensuring its centrality in international financial markets. Moreover, the technological innovations pioneered by DOGE, such as AI-driven decision-making tools and predictive analytics, can become valuable exports, opening new avenues for economic engagement with allies and partners.

Strategically, DOGE contributes to a recalibration of the U.S.'s global influence by demonstrating that governance reform is a critical component of national security. Efficient and accountable governance fosters social cohesion and economic resilience, reducing vulnerabilities that adversaries might exploit. In this way, DOGE strengthens not only the U.S.'s internal structures but also its ability to project power and support allies in an increasingly competitive global landscape.

The broader implications of DOGE extend to multilateral organizations and global governance initiatives. By aligning its principles with the goals of institutions such as the United Nations, the Organization for Economic Cooperation and Development, and the World Bank, the U.S. can advocate for the integration of efficiency and accountability into international development programs. This alignment reinforces the nation's role as a constructive partner in addressing global issues such as poverty, climate change, and technological inequality.

However, these gains are not without challenges. The visibility of DOGE's success may provoke criticism from nations that perceive U.S. leadership as a threat to their own agendas. Additionally, the scalability of DOGE's principles across diverse political and cultural contexts requires careful navigation. To mitigate these challenges, the U.S. must adopt an inclusive approach, emphasizing collaboration and mutual benefits rather than unilateralism.

Ultimately, DOGE's impact on the U.S.'s global position is a testament to the transformative power of visionary governance. By demonstrating that efficiency and accountability are not abstract ideals but achievable realities, the U.S. reaffirms its role as a beacon of progress and a partner in shaping a more equitable and effective global order. Through its leadership, the nation not only strengthens its own foundations but also contributes to a broader movement toward governance that truly serves the people.

Enhancing Trade and Alliances

Enhancing trade and alliances through the principles exemplified by DOGE provides an avenue for the United States to strengthen its global influence and economic partnerships. By embedding efficiency and accountability into its domestic governance, the U.S. not only revitalizes its internal systems but also signals to the world its renewed capacity for leadership and collaboration. This dual focus on domestic reform and international engagement positions DOGE as a catalyst for strengthening economic ties and fostering cooperative alliances.

The efficiency gains achieved through DOGE provide a foundation for reinvigorated trade relationships. By streamlining regulatory processes, improving logistical coordination, and ensuring fiscal discipline, the U.S. presents itself as a reliable and competitive partner in global markets. Reduced bureaucratic friction translates into faster and more predictable interactions with trade partners, fostering an environment conducive to mutually beneficial agreements. This operational agility strengthens the U.S.'s negotiating position, enabling it to pursue trade deals that reflect the principles of fairness, transparency, and sustainability.

Moreover, the technological advancements underpinning DOGE—such as AI-driven analytics and real-time monitoring tools—can be leveraged to enhance international trade operations. These innovations allow for more efficient customs processing, improved tracking of goods, and enhanced compliance with international standards. By sharing these capabilities with trade partners, the U.S. not only facilitates smoother exchanges but also positions itself as a leader in modernizing global trade infrastructure. Such collaborations enhance trust and deepen economic ties, creating a network of partnerships built on shared progress.

Alliances also benefit from the principles and practices demonstrated by DOGE. The emphasis on transparency and accountability resonates with allies seeking to strengthen their own governance frameworks. Collaborative initiatives, such as knowledge-sharing programs and joint reform projects, create opportunities for the U.S. to build goodwill and mutual understanding. These efforts reinforce the notion that alliances are not merely transactional but are grounded in shared values and long-term commitments to collective prosperity.

The alignment of DOGE's ethos with the goals of multilateral organizations, such as the World Trade Organization and regional trade blocs, further amplifies its impact. By advocating for efficiency and accountability in these forums, the U.S. can drive reforms that benefit the broader global community. This leadership role enhances the U.S.'s credibility and influence, ensuring its voice is central to shaping the future of international trade.

However, the benefits of enhanced trade and alliances are not confined to economic gains. Stronger economic ties foster stability and cooperation, reducing the likelihood of conflicts and creating a foundation for addressing shared challenges such as climate change, cybersecurity, and global health crises. The principles championed by DOGE, including ethical AI deployment and responsible fiscal management, contribute to a framework that balances innovation with inclusivity, ensuring that progress benefits all stakeholders.

The challenges inherent in this process are not to be underestimated. Resistance from protectionist factions, both domestically and abroad, may complicate efforts to deepen trade relationships. Additionally, the scalability of DOGE's principles in diverse international contexts requires careful adaptation to local needs and conditions. To overcome these obstacles, the U.S. must adopt a collaborative approach, emphasizing dialogue and mutual respect in all engagements.

Ultimately, the integration of DOGE's principles into trade and alliance strategies underscores the interconnected nature of governance, economics, and diplomacy. By demonstrating that efficiency and accountability are not only achievable but also transformative, the U.S. strengthens its position as a partner of choice on the global stage. Through these efforts, DOGE not only drives domestic reform but also contributes to a vision of international collaboration that is more equitable, dynamic, and resilient.

Sharing the Blueprint

Sharing the blueprint of DOGE's reforms is a transformative endeavor that extends the benefits of efficiency-driven governance beyond the boundaries of the United States. By articulating its successes and challenges in a manner that is both accessible and adaptable, DOGE enables other nations to learn from its model, fostering a global movement toward transparent, accountable, and effective administration. This dissemination of knowledge not only amplifies the impact of DOGE but also positions the United States as a leader in collaborative governance innovation.

The act of sharing a governance blueprint begins with clear documentation of principles, methodologies, and outcomes. DOGE's framework, rooted in transparency, participatory engagement, and technological integration, provides a structured foundation for others to emulate. By codifying its strategies—such as the deployment of artificial intelligence for fraud detection, the use of predictive analytics for budgeting, and the establishment of public accountability dashboards—DOGE offers a replicable guide tailored to diverse administrative contexts. Each element of this blueprint must be accompanied by case studies, data analyses, and testimonials that illustrate real-world applications and benefits.

A crucial aspect of sharing this blueprint involves fostering partnerships with international organizations and governments. Forums such as the United Nations, the Organization for Economic Cooperation and Development, and the World Bank offer platforms for dialogue and exchange, enabling DOGE to present its findings and invite collaboration. By engaging in workshops, conferences, and bilateral discussions, DOGE can build a network of reform-minded stakeholders committed to adopting and adapting its principles.

Education and training programs play a pivotal role in ensuring the successful adoption of DOGE's methodologies. Establishing training centers or offering virtual courses equips public officials from other nations with the skills and knowledge needed to implement similar reforms. These programs should focus on practical applications, emphasizing how to integrate technological tools, foster public engagement, and navigate political challenges. By empowering leaders with these resources, DOGE ensures that its impact transcends theoretical discourse, translating into actionable change.

The export of technological innovations developed under DOGE further enhances its blueprint's appeal. AI algorithms, data visualization tools, and secure digital platforms designed for transparency and accountability are not only integral to DOGE's success but also valuable assets for other governments. Licensing these technologies or forming partnerships to co-develop adaptations ensures that the blueprint includes both conceptual frameworks and tangible tools for implementation.

Recognizing the diversity of governance systems, DOGE's approach to sharing its blueprint must emphasize adaptability. What works in the U.S. may require significant adjustments to align with the political, cultural, and economic contexts of other nations. By incorporating flexibility into its recommendations, DOGE respects the sovereignty and unique needs of each government, fostering a spirit of mutual respect and cooperation.

The challenges of sharing this blueprint are as significant as its opportunities. Resistance to external influence, skepticism about scalability, and concerns over resource limitations are common obstacles. To address these, DOGE must prioritize building trust and credibility. By demonstrating transparency in its own processes and outcomes, it reassures potential adopters of the blueprint's integrity and efficacy. Engaging local leaders and stakeholders in pilot projects further mitigates resistance, providing proof of concept that aligns with local realities.

Ultimately, the sharing of DOGE's blueprint represents more than the dissemination of governance practices; it is an act of diplomacy and leadership. By positioning itself as a partner in the pursuit of global efficiency and accountability, the United States fosters a collaborative ethos that strengthens alliances and builds new bridges of understanding. Through this effort, DOGE transforms not only the mechanisms of governance but also the relationships that define our interconnected world. In doing so, it reinforces the idea that the principles of transparency, efficiency, and public empowerment are not merely national aspirations but universal values that can guide the future of governance across the globe.

Exporting Best Practices

Exporting best practices from DOGE represents an opportunity to influence global governance positively, fostering a new era of transparency, efficiency, and public trust. The principles and methodologies developed within DOGE's framework are not merely theoretical constructs but proven strategies that have demonstrated tangible benefits in streamlining government operations and enhancing accountability. Sharing these best practices with international partners allows other nations to adapt and implement similar reforms, contributing to a collective elevation of governance standards worldwide.

The foundation of exporting these practices lies in clear articulation and documentation of the strategies that underpin DOGE's success. These include the use of advanced technologies such as artificial intelligence for fraud detection and resource optimization, participatory governance models that integrate citizen feedback into decision-making processes, and transparency initiatives that provide real-time access to government operations and performance metrics. By presenting these elements in a cohesive, actionable format, DOGE creates a roadmap that other nations can follow.

International collaboration plays a pivotal role in disseminating DOGE's practices effectively. Bilateral partnerships, multilateral engagements, and participation in global forums enable the exchange of ideas and experiences, ensuring that the principles of efficiency and accountability are contextualized for diverse political and cultural environments. By engaging with organizations such as the United Nations, the World Bank, and regional alliances, DOGE facilitates the integration of its best practices into broader governance initiatives, enhancing their impact and reach.

Training and capacity-building initiatives are essential components of this effort. DOGE can establish training programs tailored to the specific needs of partner nations, equipping their policymakers, administrators, and technologists with the skills and knowledge required to implement reforms successfully. Workshops, online courses, and field visits provide immersive learning experiences, fostering a deeper understanding of how DOGE's strategies can be adapted to local contexts. These initiatives emphasize not only technical competencies but also the ethical and participatory dimensions of governance, ensuring that reforms align with democratic values.

The export of technological innovations developed under DOGE enhances its influence on global governance. Tools such as predictive analytics for budgeting, AI-driven decision-making systems, and user-friendly public accountability dashboards offer practical solutions to common challenges faced by governments worldwide. By licensing these technologies or forming public-private partnerships for their deployment, DOGE facilitates the adoption of advanced governance tools while generating economic benefits for the United States.

Customization is critical to the success of exporting best practices. Recognizing that governance systems vary widely in structure, capacity, and cultural norms, DOGE must ensure that its recommendations are flexible and adaptable. This involves working closely with partner nations to co-create solutions that address their unique challenges, fostering a sense of ownership and alignment with local priorities. Such an approach not only enhances the effectiveness of the reforms but also builds trust and goodwill between nations.

The challenges of exporting governance practices are significant but surmountable. Resistance to external influence, skepticism about the relevance of foreign models, and the logistical complexities of adapting advanced systems to developing contexts are among the hurdles that DOGE must navigate.

Addressing these requires a commitment to partnership and inclusivity, emphasizing that the transfer of best practices is a collaborative endeavor rooted in mutual respect and shared goals.

Ultimately, the export of DOGE's best practices reflects a commitment to global progress and solidarity. By sharing the lessons learned and the tools developed through its ambitious reforms, the United States contributes to a vision of governance that is more efficient, transparent, and responsive to the needs of citizens everywhere. This effort not only amplifies the legacy of DOGE but also strengthens the interconnected fabric of nations striving for a better future. Through these partnerships and exchanges, the ideals of accountability and innovation championed by DOGE transcend borders, inspiring a global transformation in the art and practice of governance.

Challenges in International Adaptation

Adapting DOGE's blueprint to diverse international contexts presents a host of challenges, underscoring the complexities of transferring governance innovations across cultural, economic, and political boundaries. While DOGE's principles of efficiency, transparency, and accountability are universally appealing, their implementation requires careful navigation of local conditions and sensitivities. These challenges highlight the necessity of a tailored approach that respects the unique characteristics of each nation while maintaining the integrity of DOGE's core ideals.

The first major challenge in adapting DOGE's model lies in the variability of governance structures. Nations differ widely in the organization of their administrative systems, the degree of centralization, and the division of powers among branches of government. DOGE's reliance on advanced technologies, streamlined workflows, and participatory mechanisms may face obstacles in settings where bureaucratic hierarchies are deeply entrenched or where digital infrastructure is underdeveloped. Addressing these disparities requires an adaptive framework that offers modular solutions, enabling governments to implement reforms incrementally and in alignment with their existing capacities.

Cultural and political contexts further complicate the transferability of DOGE's practices. Transparency and public engagement, for instance, are deeply influenced by societal norms and historical experiences. In nations where distrust of government is pervasive or where authoritarian tendencies prevail, fostering the openness and inclusivity central to DOGE's success may prove particularly challenging. Overcoming these barriers necessitates a dual strategy of education and engagement, cultivating trust through small, demonstrable successes and involving local stakeholders in the reform process from the outset.

Economic disparities also play a significant role in shaping the feasibility of adopting DOGE's innovations. Many nations, particularly in the developing world, face resource constraints that limit their ability to invest in the technological infrastructure necessary for implementing AI-driven governance or advanced data analytics. For these countries, the challenge lies in prioritizing reforms that deliver the highest impact with the lowest initial investment. International collaboration and funding, whether through multilateral organizations or bilateral partnerships, can provide critical support, enabling these nations to overcome financial barriers and access the tools needed for reform.

Resistance from vested interests poses another significant hurdle. Just as DOGE faced opposition from entrenched bureaucracies and special interest groups within the United States, international adaptations may encounter pushback from those who benefit from the status quo. Navigating this resistance requires political will and skillful diplomacy, as well as the creation of incentives that align the interests of stakeholders with the goals of reform. Demonstrating the tangible benefits of efficiency and accountability, such as improved service delivery or reduced corruption, can help to neutralize opposition and build momentum for change.

Language and communication barriers also present practical challenges in exporting DOGE's model. The technical and conceptual frameworks developed under DOGE must be translated, both literally and figuratively, to resonate with diverse audiences. This includes adapting training materials, user interfaces for technological tools, and public messaging to ensure accessibility and relevance. Collaborating with local experts and linguists can bridge these gaps, fostering a shared understanding of the principles and processes involved.

Despite these challenges, the potential for successful adaptation remains significant. By approaching international reform as a partnership rather than an imposition, DOGE can foster a spirit of collaboration that emphasizes mutual learning and shared success. Pilot projects, co-designed with host nations, provide a practical means of testing and refining DOGE's practices in new contexts, generating valuable insights that can inform broader implementation efforts. These initiatives also serve as proof points, demonstrating the adaptability and impact of DOGE's principles in diverse settings.

Ultimately, the challenges of international adaptation highlight the importance of humility, patience, and flexibility in sharing DOGE's blueprint. Recognizing that no single model fits all circumstances, DOGE must embrace the complexity of governance as an opportunity for innovation and growth. By working closely with global partners to address these challenges, the United States not only advances the principles of efficiency and accountability but also strengthens the foundations of international cooperation and progress. Through these efforts, DOGE transforms from a national initiative into a global movement, inspiring a new era of governance that serves the needs of all people.

Chapter 10: The Road Ahead

Scaling DOGE's Success

Scaling the success of DOGE from a pioneering federal initiative to a universally applicable framework represents the next logical step in its journey. The principles that have driven its achievements—efficiency, transparency, and accountability—can serve as a foundation for reforms at every level of governance. Expanding DOGE's reach involves not only replicating its strategies across diverse federal agencies but also adapting them for state, local, and even international applications. This effort is both an opportunity to amplify its impact and a challenge that requires careful planning, collaboration, and innovation.

At the heart of scaling DOGE's success is the need to maintain its core philosophy while allowing for contextual flexibility. The federal government's complexity provided the perfect testing ground for DOGE's methods, showcasing how advanced technologies, participatory governance, and strategic fiscal planning can transform sprawling bureaucracies. Applying these lessons to smaller or differently structured organizations requires a nuanced understanding of their unique challenges and opportunities. For instance, state governments often operate with fewer resources but have closer relationships with their constituents, creating opportunities for more direct citizen engagement and rapid policy implementation.

Technology plays a central role in enabling this scalability. The AI-driven tools and data analytics systems developed under DOGE can be tailored to suit the needs of smaller entities or organizations with different mandates. Predictive budgeting models, fraud detection algorithms, and public accountability dashboards are not only transferable but also inherently adaptable, capable of being scaled up or down based on the size and scope of the institution. By creating modular versions of these tools, DOGE ensures that they can be deployed effectively across various levels of governance.

Building partnerships is another critical component of scaling success. Collaboration with state and local governments, as well as with non-governmental organizations and private-sector entities, fosters the exchange of knowledge and resources needed to implement reforms. These partnerships allow DOGE's principles to be integrated into existing structures without duplicating efforts or overburdening smaller institutions. Pilot projects can serve as proving grounds for these adaptations, demonstrating their feasibility and effectiveness in real-world scenarios.

Expanding DOGE's reach also requires investment in training and capacity building. Scaling success is not simply about introducing new tools or protocols; it involves equipping leaders and administrators with the skills and mindset needed to embrace change. Training programs tailored to the specific needs of state and local officials ensure that they can confidently implement DOGE-inspired reforms. These programs should emphasize both technical skills and the ethical dimensions of governance, ensuring that efficiency does not come at the expense of equity or accountability.

As DOGE's principles spread, it is essential to address potential challenges proactively. Resistance to change, whether due to political opposition, resource constraints, or institutional inertia, remains a significant obstacle. Overcoming this requires clear communication of the benefits of reform, backed by data and success stories that illustrate tangible improvements. Engaging communities in the reform process also helps to build trust and momentum, ensuring that changes are supported by those they affect most directly.

The success of scaling DOGE depends on balancing innovation with continuity. While the initiative's adaptability is one of its greatest strengths, its expansion must remain true to the values and goals that have defined its success. Transparency, citizen empowerment, and fiscal responsibility are not merely strategies—they are guiding principles that must inform every decision and adaptation.

Ultimately, scaling DOGE's success represents a broader vision for governance. It demonstrates that efficiency and accountability are not confined to individual initiatives but are part of a larger movement toward a more responsive and effective government. By extending its reach, DOGE not only reinforces its legacy but also contributes to a nationwide—and potentially global—culture of innovation and reform. Through this effort, the principles of DOGE become more than a blueprint for change; they become the foundation for a government that serves all people with integrity and purpose.

Expanding Reform Beyond Federal Agencies

Expanding reform beyond the federal agencies that have served as DOGE's primary focus requires an understanding of the broader landscape of governance in the United States. The federal system, while vast and complex, is interconnected with a web of state and local governments, each facing unique challenges but also sharing many of the inefficiencies and redundancies that DOGE has sought to address. Extending the principles of transparency, accountability, and efficiency into these levels of government represents an ambitious but achievable goal, one that promises to amplify the transformative impact of DOGE's initiatives.

State and local governments, while often constrained by limited resources, are positioned to implement reforms with agility and precision. These governments interact directly with citizens in delivering critical services, from education and public safety to infrastructure maintenance and health care. Their proximity to the populations they serve allows for more immediate feedback and adjustments, making them ideal candidates for testing and adapting DOGE's methodologies. By leveraging the lessons learned at the federal level, state and local agencies can enhance service delivery, reduce costs, and rebuild public trust.

The integration of DOGE-inspired reforms begins with identifying areas of overlap and inefficiency across various levels of government. Redundancies between state and federal programs, for instance, often result in unnecessary expenditure and confusion for beneficiaries. Streamlining these interactions through better coordination and shared technology platforms can significantly enhance efficiency. Similarly, adopting AI-driven tools for fraud detection and predictive analytics in state budget planning mirrors the successes seen in federal agencies, demonstrating the scalability of these innovations.

Training and capacity building play a critical role in empowering state and local governments to adopt these reforms. Many smaller jurisdictions may lack the expertise or infrastructure needed to implement advanced governance tools. DOGE can facilitate this transition by providing access to training programs, technical assistance, and shared resources. Initiatives such as intergovernmental partnerships and mentorship programs enable knowledge transfer and foster a culture of collaboration, ensuring that reforms are not only implemented but sustained.

Transparency and public engagement remain central to the success of expanding reforms. Citizens must see and understand the benefits of these initiatives to trust and support them. Public accountability dashboards, participatory budgeting platforms, and regular town hall discussions provide opportunities for community members to engage directly with the reform process. By involving citizens in shaping priorities and monitoring progress, state and local governments can build the same trust and momentum that DOGE has cultivated at the federal level.

Scaling reforms to state and local levels also presents opportunities for innovation tailored to specific regional needs. States with large agricultural economies, for example, could use DOGE's data analytics models to optimize resource allocation in farming subsidies and environmental conservation programs. Urban centers grappling with housing crises might adapt DOGE's predictive analytics tools to forecast demand and streamline housing policies. These localized applications not only demonstrate the versatility of DOGE's principles but also highlight their potential to address diverse challenges across the nation.

However, expanding reform efforts is not without its challenges. Resistance from entrenched interests, resource disparities between states, and political fragmentation can impede progress. To address these obstacles, DOGE must emphasize collaboration and inclusivity. Engaging a broad coalition of stakeholders, from public officials and community leaders to private-sector partners, ensures that reforms are implemented in a manner that respects local contexts while adhering to the overarching principles of efficiency and accountability.

Ultimately, extending DOGE's success beyond federal agencies is about creating a cohesive framework for governance reform that transcends levels of government. By empowering state and local entities to adopt and adapt these principles, the United States can create a more integrated and responsive system of governance. This effort not only reinforces the achievements of DOGE but also lays the foundation for a nation where every level of government operates with the transparency, efficiency, and effectiveness that citizens deserve. Through this expansion, the vision of DOGE becomes not just a federal initiative but a national movement, one that redefines the relationship between government and the people it serves.

Long-Term Goals for Governance

Establishing long-term goals for governance reform is essential to ensuring that the principles and successes of DOGE endure beyond the immediate achievements of its initiatives. These goals must transcend short-term metrics and focus on fostering a culture of innovation, accountability, and adaptability within government institutions. By envisioning a future where efficiency and transparency are not merely strategies but foundational values, DOGE sets the stage for sustained transformation that can weather political shifts, economic uncertainties, and evolving societal needs.

A primary long-term objective of governance reform is institutionalizing efficiency as a core operational philosophy. This involves embedding systems and processes that continually evaluate performance, identify inefficiencies, and implement improvements. Tools such as artificial intelligence, predictive analytics, and real-time data monitoring should not only be adopted but also regularly updated to align with technological advancements. Establishing frameworks for ongoing evaluation ensures that government agencies remain agile and responsive to changing demands, preventing the stagnation that often accompanies bureaucratic systems.

Transparency, as a pillar of governance, must evolve from an initiative into an ingrained expectation. Long-term goals should prioritize the development of mechanisms that make information access seamless and intuitive for citizens. Public accountability dashboards, for instance, should expand to include deeper insights into government spending, project timelines, and policy impacts. These platforms must be designed with accessibility in mind, leveraging advances in user interface design and data visualization to engage a diverse audience. By fostering a culture where transparency is not only practiced but celebrated, government can strengthen public trust and encourage greater civic participation.

Another critical goal is enhancing the adaptability of governance structures to future challenges. The world is becoming increasingly interconnected and unpredictable, with technological disruptions, climate change, and global economic shifts demanding flexibility in policymaking and administration. DOGE's framework must include mechanisms for scenario planning and rapid response, allowing agencies to pivot effectively in the face of emerging crises. This adaptability also requires fostering a workforce skilled in change management, problem-solving, and innovative thinking.

Building a governance ecosystem that prioritizes equity alongside efficiency is equally important. As reforms optimize processes and cut costs, care must be taken to ensure that vulnerable populations are not disproportionately affected. Long-term goals should emphasize inclusive policy design, integrating diverse perspectives into decision-making processes. This includes expanding public engagement initiatives, such as town halls and online forums, to reach underrepresented communities and incorporating their input into governance strategies.

Fiscal sustainability remains a cornerstone of long-term governance goals. Beyond reducing waste and optimizing budgets, DOGE must advocate for systemic changes in fiscal policy that promote responsible spending and investment. Establishing multi-year budgeting cycles and integrating comprehensive cost-benefit analyses into policymaking processes ensure that resources are allocated efficiently and effectively. These measures not only enhance fiscal discipline but also provide a clear roadmap for achieving financial stability while supporting essential services.

Leadership development is another critical component of ensuring lasting reform. Government leaders at all levels must be equipped with the skills and knowledge needed to champion efficiency and transparency. Investing in leadership training programs that emphasize ethical governance, data-driven decision-making, and public engagement creates a pipeline of officials capable of sustaining and advancing DOGE's vision. Partnerships with academic institutions and think tanks can provide additional support, fostering a culture of continuous learning and innovation within the public sector.

To anchor these long-term goals, DOGE must establish benchmarks and metrics that track progress over time. These indicators should be designed to measure not only operational efficiency and cost savings but also the broader societal impacts of reform, such as improvements in public trust, service delivery, and economic opportunity. Regular reporting on these metrics reinforces accountability and ensures that reforms remain aligned with their intended outcomes.

The pursuit of long-term governance goals is not a finite project but a perpetual journey. By laying the groundwork for sustained innovation, accountability, and inclusivity, DOGE transforms the concept of reform from a series of discrete initiatives into a continuous process of improvement. This vision ensures that government institutions are not only prepared to meet the challenges of today but are also equipped to thrive in the dynamic and uncertain landscape of the future. Through these efforts, DOGE secures its legacy as a catalyst for a new era of governance that serves the people with integrity and purpose.

Preparing for the Unexpected

Preparing for the unexpected is both a philosophical and practical imperative for governance in an era marked by rapid technological evolution, global interconnectivity, and unprecedented challenges. DOGE's framework, rooted in efficiency and adaptability, must evolve further to encompass mechanisms that anticipate and respond to uncertainty. By embedding resilience into every layer of governance, DOGE sets a precedent for a government that is not only efficient but also agile in the face of change.

At its core, readiness for the unexpected demands a shift in mindset from reactive to proactive governance. This begins with the integration of scenario planning into policy development. Leveraging data analytics and predictive modeling, DOGE can simulate a range of potential futures, exploring how various disruptions—be they economic, environmental, or geopolitical— might impact government operations and public services. These insights enable policymakers to craft strategies that are robust across multiple scenarios, ensuring that the government remains effective even in times of crisis.

Flexibility in resource allocation is another critical component of preparedness. Traditional budgeting processes often lock resources into rigid categories, limiting the government's ability to pivot when priorities shift. By adopting dynamic budgeting systems informed by real-time data, DOGE can reallocate funds swiftly and strategically, directing them toward emergent needs without undermining ongoing programs. This approach not only enhances efficiency but also builds a financial buffer against unforeseen demands.

Technology plays a pivotal role in preparing for the unexpected. The AI-driven tools and digital platforms employed by DOGE can be further enhanced to include early warning systems that detect anomalies and forecast potential disruptions. For instance, real-time monitoring of economic indicators could signal looming financial crises, while advanced analytics could identify vulnerabilities in supply chains or infrastructure. By equipping agencies with these capabilities, DOGE ensures that responses to challenges are timely and informed.

Human capital is equally essential in building resilience. A workforce trained to adapt to change, innovate solutions, and collaborate across disciplines is better equipped to navigate uncertainty. DOGE must prioritize professional development programs that cultivate these skills, fostering a culture of continuous learning and problem-solving. Additionally, establishing cross-agency task forces that can be mobilized quickly during crises ensures that expertise and resources are coordinated effectively.

Transparency and communication are indispensable in managing the unexpected. Citizens must understand not only what the government is doing to address challenges but also why these actions are necessary. DOGE's commitment to public engagement can be expanded to include crisis communication strategies that provide clear, consistent, and accessible information. This transparency builds trust and mitigates the fear and misinformation that often accompany unexpected events.

While preparation is key, no strategy can eliminate uncertainty entirely. DOGE must therefore embrace the principle of iterative improvement, using each crisis as an opportunity to refine its systems and strategies. After-action reviews and feedback loops allow for the identification of strengths and weaknesses in the government's response, informing future policies and practices. This commitment to learning ensures that the government becomes stronger and more capable with each challenge it faces.

Preparing for the unexpected also involves fostering collaboration with external partners. No single government can address the complexities of today's interconnected world alone. DOGE's approach should include building alliances with other nations, international organizations, and private-sector innovators. These partnerships enable the sharing of resources, expertise, and best practices, creating a network of resilience that extends beyond national borders.

Ultimately, readiness for the unexpected is about embedding resilience into the DNA of governance. By anticipating challenges, fostering flexibility, and prioritizing continuous improvement, DOGE redefines what it means to govern in an age of uncertainty. This vision of preparedness not only enhances the government's capacity to serve its citizens but also inspires confidence in its ability to navigate whatever the future may hold. Through these efforts, DOGE transforms governance into a dynamic, responsive, and enduring institution that stands ready to meet the demands of an ever-changing world.

Adapting to Future Challenges

Adapting to future challenges is a vital cornerstone of governance reform, particularly in a world characterized by rapid technological evolution, environmental unpredictability, and shifting geopolitical landscapes. For DOGE, the ability to foresee, prepare for, and navigate these challenges is essential to maintaining its role as a dynamic force for efficiency and accountability. This adaptability requires a forward-looking strategy that integrates technological innovation, flexible policy design, and a commitment to ethical governance.

One of the most pressing future challenges lies in the continued integration of emerging technologies into government operations. Artificial intelligence, quantum computing, and blockchain hold immense potential to revolutionize governance, yet they also present significant risks. DOGE must remain at the forefront of technological adoption while establishing robust ethical frameworks to address concerns such as algorithmic bias, data security, and the potential for technology to exacerbate inequality. By fostering partnerships with leading innovators and academic institutions, DOGE can ensure that its technological tools are both cutting-edge and responsibly developed.

Climate change represents another critical area where adaptability is paramount. The increasing frequency of extreme weather events and the long-term effects of global warming demand a governance model that is not only efficient but also resilient. DOGE's strategies should include integrating environmental forecasting into planning processes, enhancing disaster response capabilities, and promoting sustainable practices across all government operations. This includes leveraging AI and data analytics to optimize resource allocation for disaster mitigation and recovery efforts, ensuring that responses are swift, effective, and equitable.

Economic volatility is yet another dimension of the future that requires proactive governance. Fluctuating markets, trade disruptions, and the potential for financial crises necessitate a framework that prioritizes fiscal responsibility while maintaining flexibility. DOGE's use of predictive analytics for budgetary planning must evolve to incorporate scenario modeling that accounts for global economic shifts. This enables policymakers to craft budgets that are not only balanced but also robust against external shocks, preserving critical services even in times of financial strain.

Geopolitical challenges, including the shifting dynamics of international relations and the rise of non-state actors, demand an agile and informed approach to governance. DOGE must work closely with the Department of State and other agencies to ensure that its strategies align with national security objectives and global stability. By incorporating intelligence-driven insights into its operations, DOGE can anticipate and mitigate risks, contributing to a more secure and resilient government.

Public trust and civic engagement are foundational to addressing future challenges. As the landscape of governance evolves, citizens must remain active participants in shaping policies and priorities. DOGE's initiatives to enhance transparency and accountability must continue to expand, incorporating new platforms and tools that facilitate meaningful engagement. Ensuring that reforms are inclusive and responsive to diverse perspectives is critical to maintaining legitimacy and public support in an increasingly interconnected and informed society.

The unpredictability of future challenges underscores the need for continuous learning and improvement within DOGE. Establishing a culture of experimentation and feedback allows for the refinement of policies and practices based on real-world outcomes. Pilot programs, iterative design, and regular evaluations ensure that DOGE remains adaptable and effective, even as the nature of challenges evolves.

Ultimately, preparing for the future is not just about addressing potential crises; it is about embracing change as an opportunity to innovate and improve. For DOGE, this means integrating foresight into every aspect of its operations, from technology adoption and fiscal planning to public engagement and environmental stewardship. Through this commitment to adaptability, DOGE not only positions itself as a model for effective governance but also ensures that it is ready to meet the demands of an uncertain and dynamic world.

Continuous Improvement Strategies

Continuous improvement strategies are essential for ensuring that governance remains a living, evolving system capable of meeting the needs of citizens in a dynamic and unpredictable world. For DOGE, this principle is foundational, emphasizing not only the importance of achieving immediate efficiency gains but also the necessity of embedding adaptability and innovation into the DNA of government operations. Continuous improvement is not a destination but a process—an ongoing commitment to excellence that requires the active engagement of technology, people, and processes.

The cornerstone of continuous improvement lies in creating feedback loops that inform policy refinement and operational adjustments. Real-time data analytics, a hallmark of DOGE's technological approach, enables the government to monitor performance metrics across agencies, identifying areas of success and pinpointing inefficiencies. This data-driven insight provides the foundation for iterative improvement, ensuring that strategies can be fine-tuned based on measurable outcomes rather than assumptions. By institutionalizing mechanisms for regular evaluation, such as performance reviews and public audits, DOGE ensures that progress is not static but constantly advancing.

Technology plays a pivotal role in facilitating this ongoing refinement. Artificial intelligence, machine learning, and predictive modeling offer powerful tools for identifying patterns, forecasting challenges, and proposing solutions. These technologies can automate the detection of inefficiencies, such as redundant processes or misaligned resource allocations, allowing agencies to act swiftly and decisively. Additionally, AI-driven simulations enable policymakers to test the potential impacts of proposed reforms in a controlled environment, minimizing risks while maximizing the likelihood of success.

Equally important is fostering a culture of innovation within government agencies. Continuous improvement thrives in an environment where creativity and experimentation are encouraged, and where failures are viewed not as setbacks but as opportunities for learning. DOGE must prioritize initiatives that empower employees at all levels to contribute ideas, challenge existing paradigms, and experiment with new approaches. Creating innovation labs within agencies, for instance, provides a structured space for testing and developing novel solutions without disrupting day-to-day operations.

Training and professional development are integral to sustaining this culture of improvement. A workforce equipped with the skills to analyze data, implement technological solutions, and manage change is better prepared to adapt to evolving demands. DOGE's investment in capacity building ensures that employees remain at the forefront of governance innovation, driving progress from within. This emphasis on human capital also extends to leadership development, as visionary and adaptive leaders are critical to championing and sustaining reform.

Public engagement is another essential element of continuous improvement. Citizens are not just beneficiaries of governance but active participants whose insights and experiences can drive meaningful change. By expanding platforms for citizen feedback, such as online portals and town hall forums, DOGE ensures that reforms are responsive to the needs and priorities of the public. Transparency in decision-making processes further strengthens this relationship, fostering trust and collaboration between the government and its constituents.

While the tools and principles of continuous improvement are powerful, their success depends on overcoming potential obstacles. Resistance to change, whether from entrenched interests or institutional inertia, remains a significant challenge. To address this, DOGE must focus on building consensus and aligning incentives, demonstrating through evidence and communication the tangible benefits of reform. Clear articulation of the vision and progress of continuous improvement efforts ensures that stakeholders remain engaged and supportive.

In the long term, the sustainability of continuous improvement requires embedding it into the structural framework of governance. Establishing dedicated offices or committees focused on innovation and reform ensures that this commitment endures across administrations and political cycles. These entities act as stewards of progress, maintaining momentum and safeguarding against regression.

Ultimately, continuous improvement strategies are a testament to DOGE's philosophy of governance as a living system—one that grows, learns, and evolves in response to the world around it. By embracing this principle, DOGE not only transforms the operations of government but also redefines its relationship with the people it serves. In this vision, governance is not a static institution but a dynamic partnership, rooted in the shared pursuit of a better, more efficient, and more responsive society. Through these strategies, DOGE ensures that its impact is not only enduring but also ever-expanding, setting a standard of excellence for generations to come.

Legacy and Vision

The legacy of DOGE represents more than a series of reforms; it encapsulates a vision of governance that is leaner, more transparent, and fundamentally aligned with the needs of the people it serves. As the United States navigates a pivotal chapter in its history, the achievements of DOGE stand as a testament to what is possible when bold ideas meet rigorous implementation. The enduring impact of this initiative lies not only in the immediate efficiency gains it delivers but also in the broader cultural shift it inspires—a shift toward accountability, innovation, and public trust.

DOGE's legacy begins with its tangible successes: billions saved through streamlined processes, waste eliminated across federal agencies, and a renewed focus on results-driven governance. Yet, its true significance extends far beyond these metrics. By demonstrating that government can operate with the agility and foresight often associated with the private sector, DOGE redefines the very concept of public administration. It shows that efficiency is not anathema to integrity or inclusivity but a catalyst for achieving both at scale.

The visionary aspect of DOGE lies in its ability to anticipate and adapt to future challenges. The incorporation of artificial intelligence, advanced analytics, and participatory platforms not only addresses present inefficiencies but also lays the groundwork for a government that evolves alongside societal and technological progress. This foresight ensures that DOGE is not merely a reaction to past failures but a proactive model for enduring excellence.

Central to this vision is the democratization of governance. DOGE empowers citizens to play an active role in shaping policies and monitoring outcomes, transforming the relationship between government and the governed. Initiatives such as public accountability dashboards and participatory budgeting create a sense of shared ownership, ensuring that reforms are not only sustainable but also deeply rooted in the collective will of the people. This participatory ethos cements DOGE's legacy as a movement that transcends bureaucracy to embody the principles of democracy itself.

As DOGE continues to influence governance at all levels, its vision serves as a guidepost for future reforms. The principles of efficiency and accountability it champions are not confined to the federal government; they are universally applicable, offering a blueprint for state and local administrations, private institutions, and even international bodies. By fostering partnerships and sharing best practices, DOGE extends its impact globally, demonstrating that the challenges of governance are shared and that their solutions require collective effort.

Yet, the ultimate measure of DOGE's legacy lies in its ability to inspire a lasting commitment to improvement. It is not enough for reforms to achieve immediate success; they must also instill a culture of continuous evaluation and adaptation. DOGE's framework of iterative progress ensures that government remains a living system, capable of learning from its successes and failures alike. This emphasis on resilience and growth positions

DOGE not just as a response to a moment of crisis but as a foundation for a more responsive and equitable future.

The vision articulated by DOGE challenges us to imagine governance not as a static institution but as a dynamic partnership—one that thrives on transparency, inclusivity, and innovation. It calls upon policymakers, administrators, and citizens alike to embrace the possibilities of reform and to recognize that the pursuit of efficiency is inseparable from the pursuit of justice and equity. In doing so, DOGE transforms governance into a shared endeavor, rooted in the belief that government exists to serve the people and that its effectiveness is measured by the trust and well-being of its citizens.

Through its legacy and vision, DOGE redefines what is possible in governance, offering a model that is both aspirational and attainable. It stands as a reminder that meaningful change begins with bold ideas and is sustained by unwavering commitment. As the United States looks to the future, the lessons of DOGE will remain a guiding light, illuminating the path toward a government that is truly of, by, and for the people.

DOGE's Place in American History

DOGE's place in American history is cemented as a transformative force, a bold experiment in reimagining governance that aligns with the nation's highest ideals of efficiency, transparency, and accountability. Against the backdrop of a federal system often criticized for its inertia and wastefulness, DOGE emerged as a beacon of innovation, reshaping the narrative around public administration and setting a new standard for what government can achieve. Its legacy, both in its successes and the principles it champions, stands as a defining moment in the ongoing evolution of the United States' democratic experiment.

What distinguishes DOGE in the annals of American history is its synthesis of technological innovation and constitutional values. The initiative drew upon the entrepreneurial ingenuity of its leadership and the rich traditions of participatory governance, creating a model that was at once cutting-edge and deeply rooted in the principles of the republic. By leveraging tools such as artificial intelligence and data-driven decision-making, DOGE exemplified the potential of technology to enhance—not replace—human judgment and accountability in governance. This balance between innovation and ethical stewardship represents one of DOGE's most enduring contributions to the American ethos.

The historical significance of DOGE also lies in its ability to respond to a critical moment of national need. Faced with mounting debt, public disillusionment, and growing skepticism about the effectiveness of federal institutions, DOGE offered a path forward that was both pragmatic and visionary. Its strategies for eliminating waste, streamlining processes, and enhancing transparency resonated not only as fiscal imperatives but also as moral commitments to a government that serves its people efficiently and equitably. In doing so, DOGE restored a measure of trust in public institutions, a trust that is essential to the functioning of a healthy democracy.

DOGE's influence extends beyond the immediate context of its implementation. It serves as a case study in the power of leadership to drive change. The appointment of figures such as Elon Musk and Vivek Ramaswamy to key roles underscored the importance of visionary thinking and interdisciplinary expertise in tackling complex challenges. Their leadership demonstrated how entrepreneurial principles—innovation, agility, and a results-oriented mindset—can be applied effectively within the public sector, challenging the conventional wisdom that bureaucracy is inherently resistant to change.

The cultural impact of DOGE is equally profound. By introducing participatory tools like public accountability dashboards and crowdsourced policy initiatives, DOGE redefined the relationship between government and citizens. It showed that transparency is not merely a policy goal but a dynamic process that fosters civic engagement and empowers individuals to take an active role in governance. This shift toward a more interactive and inclusive model of public administration reflects a broader trend in American society toward decentralization and collaborative problem-solving, trends that DOGE helped to accelerate and legitimize.

Historically, DOGE stands alongside other pivotal reform efforts, such as the Grace Commission of the 1980s, but it distinguishes itself through its embrace of technology and its holistic approach to governance. While past initiatives often focused narrowly on cost-cutting or regulatory streamlining, DOGE addressed the structural and cultural dimensions of inefficiency, offering a comprehensive blueprint for reform. This breadth of vision ensures that DOGE will be remembered not only for its practical achievements but also for its ambitious reimagining of what government can be.

In the broader arc of American history, DOGE represents a reaffirmation of the nation's ability to adapt and innovate in the face of challenges. It echoes the spirit of other transformative periods, from the New Deal to the space race, when the United States harnessed its resources, creativity, and determination to overcome obstacles and redefine its future. Like these milestones, DOGE serves as a reminder that the promise of democracy lies in its capacity for renewal—a capacity that depends on bold ideas, courageous leadership, and the collective will to make them a reality.

As historians and policymakers reflect on DOGE's legacy, they will see a moment when the United States chose to confront its inefficiencies not with incremental adjustments but with a wholesale commitment to reinvention. This choice, rooted in both pragmatism and principle, ensures that DOGE's place in American history is not only secure but also instructive, offering lessons for future generations about the enduring power of reform and the infinite potential of a government that truly serves its people.

The Nation's Path Forward

The nation's path forward, as illuminated by the transformative principles of DOGE, is one of renewed purpose and enduring commitment to governance that truly serves its people. This vision is not limited to the immediate gains of streamlined processes or fiscal prudence but extends to a broader redefinition of how government can operate in a modern, interconnected world. Moving forward, the legacy of DOGE offers a framework for sustained progress, innovation, and resilience, anchoring the United States firmly as a model of effective democracy.

A critical aspect of the nation's future lies in institutionalizing the reforms initiated under DOGE. This involves embedding the principles of efficiency, transparency, and accountability into the structural foundation of governance. Agencies must evolve from adopting individual reforms to fostering a culture where continuous improvement is second nature. This shift ensures that efficiency is not an episodic goal but a permanent expectation, enabling government operations to remain agile and responsive.

To sustain the momentum of change, the nation must invest in the technologies that powered DOGE's success. Artificial intelligence, predictive analytics, and blockchain solutions have proven instrumental in reducing waste and improving decision-making. The path forward includes not only maintaining these systems but also scaling and innovating upon them to address emerging challenges. This requires ongoing collaboration with the private sector and academia, fostering a dynamic ecosystem of technological advancement that supports public administration.

Public engagement will remain a cornerstone of progress. The participatory governance models championed by DOGE have demonstrated the value of citizen involvement in shaping policies and monitoring outcomes. Moving forward, the government must expand these platforms, ensuring that they are accessible to all segments of society. By creating spaces where diverse voices can contribute to the policymaking process, the government reinforces trust and inclusivity, critical components of a thriving democracy.

Fiscal sustainability continues to be a vital priority. The lessons learned from DOGE's budgetary reforms must inform a long-term strategy for managing the nation's resources responsibly. This involves not only eliminating waste but also making strategic investments in infrastructure, education, and healthcare, ensuring that economic growth is both robust and equitable. Balancing immediate fiscal needs with future aspirations is essential to maintaining economic stability and global competitiveness.

On the international stage, the principles of DOGE offer an opportunity for the United States to lead by example. By sharing best practices and fostering global partnerships, the nation can inspire similar reforms in other countries, promoting a shared commitment to efficiency and transparency. This collaborative approach enhances diplomatic relations and underscores the United States' role as a leader in governance innovation.

The path forward also requires addressing the challenges of tomorrow. Climate change, technological disruptions, and evolving geopolitical landscapes demand a government that is not only efficient but also resilient. Building on DOGE's foundation, the nation must prioritize adaptability, ensuring that its systems and policies can withstand and respond to unforeseen events. This includes fostering a workforce skilled in problem-solving and innovation, capable of navigating complex and rapidly changing environments.

Ultimately, the nation's path forward is guided by a profound belief in the power of governance to effect positive change. DOGE has shown that even in the face of entrenched inefficiencies and skepticism, bold ideas and steadfast leadership can transform institutions. As the United States embraces this vision, it reaffirms its commitment to a government that reflects the highest aspirations of its people—one that is efficient, equitable, and deeply attuned to the needs of a changing world.

Through the continued pursuit of these principles, the nation not only honors the legacy of DOGE but also charts a course for a future where governance is a dynamic and empowering force for all. This journey, fueled by innovation, collaboration, and a relentless drive for improvement, ensures that the promise of democracy remains vibrant and enduring, a beacon for generations to come.

Conclusion: A Nation Reborn

Reflection on the Journey

Reflecting on the journey of DOGE is to acknowledge a profound transformation—a movement that reshaped the foundational tenets of governance, brought efficiency to the fore, and rekindled a sense of trust in public institutions. It is a journey marked by bold decisions, unwavering resolve, and a vision that transcended the challenges of its time. To look back is to see not only what was achieved but also the values that drove those achievements: transparency, accountability, and a commitment to serving the people.

The origins of DOGE, rooted in a critical moment of national reckoning, remind us of the urgency that spurred its creation. Facing mounting inefficiencies and a ballooning national debt, the government was at a crossroads. DOGE emerged as a response to this crisis, championing a new model of governance that emphasized the elimination of waste, the integration of technology, and the empowerment of citizens. The challenges were immense, but so too was the potential for transformation.

Throughout its journey, DOGE demonstrated the power of innovation to overcome entrenched inertia. The deployment of artificial intelligence to analyze budgets, detect fraud, and streamline operations revealed the possibilities of technology as a tool for accountability. These advancements did not just address inefficiencies; they redefined what government could achieve when guided by data-driven decision-making and ethical leadership. The introduction of participatory platforms further underscored DOGE's commitment to inclusivity, fostering a dialogue between the government and its citizens that was unprecedented in its depth and transparency.

The leadership of DOGE, exemplified by figures like Elon Musk and Vivek Ramaswamy, played a pivotal role in steering this transformation. Their ability to meld entrepreneurial agility with public sector challenges set a new standard for leadership in governance. It was their vision that ensured DOGE remained focused on its core mission while navigating the complexities of political resistance, public skepticism, and logistical hurdles. Their work highlighted the importance of interdisciplinary collaboration and the potential for private sector expertise to complement public service goals.

As the initiatives of DOGE unfolded, the tangible impacts became clear. Billions in savings, streamlined processes, and enhanced public engagement were not just outcomes—they were proof that governance could be reimagined. Yet, the journey was not without its struggles. Balancing efficiency with equity, addressing fears of job displacement, and ensuring the ethical deployment of technology required careful deliberation and a steadfast adherence to core democratic principles.

Reflecting on the lessons of DOGE, it becomes evident that its greatest achievement lies not in the numbers it produced but in the cultural shift it initiated. It rekindled the belief that government could be a force for good, capable of adapting to the needs of its people and the demands of its time. It demonstrated that the pursuit of efficiency is not a departure from democratic ideals but a realization of them, aligning governance with the values of fairness, transparency, and accountability.

The journey of DOGE is not merely a chapter in the history of governance; it is a roadmap for the future. It invites us to imagine what is possible when courage and innovation converge, when leaders dare to challenge the status quo, and when citizens are empowered to shape their government. As we reflect, we also look forward, inspired by the legacy of DOGE to continue the work of building a government that serves all with integrity and purpose.

In this reflection, we find a narrative of transformation, resilience, and hope—a reminder that the path of progress, though arduous, is always worth pursuing. DOGE stands as a testament to what can be achieved when vision meets action, leaving an indelible mark on the fabric of American governance. Its journey, though rooted in its time, is timeless in its lessons, offering guidance and inspiration for generations to come.

From Crisis to Transformation

From crisis to transformation, DOGE's journey exemplifies the resilience of a nation willing to confront its most entrenched challenges with bold vision and decisive action. The crisis that precipitated DOGE's inception was not merely a fiscal reckoning but a fundamental test of governance. Years of bureaucratic stagnation, inefficiency, and escalating debt had created a government that struggled to meet the demands of its citizens. Against this backdrop, DOGE emerged as both a response and a revolution, reshaping the very architecture of federal administration and reaffirming the principles of accountability and innovation.

The early days of DOGE were defined by urgency. The national debt, having reached unprecedented levels, posed an existential threat to economic stability and global standing. Wasteful spending, redundant processes, and opaque decision-making had eroded public trust and created an unsustainable status quo. It was clear that incremental reforms would no longer suffice; what was needed was a systemic overhaul. DOGE's mandate—to cut inefficiencies, enhance transparency, and modernize governance—was ambitious, but ambition was the only path forward.

At the heart of this transformation was a commitment to leveraging technology as a catalyst for change. The use of artificial intelligence to analyze data, detect inefficiencies, and optimize resource allocation proved to be a game-changer. By automating mundane tasks and identifying patterns invisible to human analysts, AI allowed DOGE to tackle inefficiencies at a scale and speed previously unimaginable. But technology was only part of the equation. Equally important was the human element: the leadership, vision, and collaboration that turned abstract ideas into actionable strategies.

Under the stewardship of leaders like Elon Musk and Vivek Ramaswamy, DOGE exemplified the power of interdisciplinary thinking. Musk's technological foresight and Ramaswamy's focus on public engagement created a dynamic synergy that guided DOGE through its formative challenges. Their leadership demonstrated that efficiency and accountability are not mutually exclusive but are, in fact, mutually reinforcing. By fostering a culture of transparency and inviting public participation, DOGE not only streamlined operations but also rebuilt trust between government and citizens.

The transformation wrought by DOGE was not without its challenges. Resistance from entrenched interests, fears of job displacement, and the inherent inertia of bureaucracy all posed significant obstacles. Yet, these challenges also underscored the necessity of reform. They revealed the depth of the inefficiencies that had taken root and the urgency of addressing them. Through perseverance and a commitment to its core principles, DOGE overcame these barriers, proving that meaningful change is possible even in the most complex systems.

As the reforms took hold, the impact was profound. Billions of dollars in savings were realized, public services became more efficient and accessible, and a new standard for governance was established. These achievements were not merely practical successes; they were symbolic of a deeper renewal. They demonstrated that a government willing to innovate, adapt, and listen to its citizens could rise to meet the challenges of the 21st century.

The transformation initiated by DOGE also had a ripple effect beyond the federal government. State and local administrations began adopting similar strategies, inspired by DOGE's success. Internationally, DOGE became a model for other nations grappling with their own governance challenges, showcasing the universal applicability of its principles. This broader impact underscores the transformative power of ideas when coupled with effective implementation.

Reflecting on this journey, it is clear that DOGE was not just a response to a crisis but a reinvention of what governance can and should be. It redefined efficiency not as a narrow goal but as a holistic philosophy that encompasses fiscal responsibility, technological innovation, and ethical stewardship. From its inception to its enduring legacy, DOGE represents a profound shift—a testament to the resilience and creativity of a nation determined to thrive. As the United States continues to build on this foundation, the lessons of DOGE will remain a guiding light, illuminating the path from crisis to enduring transformation.

The Promise of Efficiency

The promise of efficiency lies at the heart of DOGE's transformative journey, offering a vision of governance that is not only functional but inspiring in its capacity to uplift the nation. Efficiency, often mischaracterized as a sterile metric or mere cost-cutting measure, takes on a profound significance when viewed through the lens of DOGE's achievements. It becomes a philosophy of governance—a commitment to maximizing value, reducing waste, and ensuring that every action taken by the government serves a meaningful purpose.

The concept of efficiency as realized by DOGE transcends the simple elimination of redundancies. It embodies the art of creating systems that are responsive, adaptable, and equitable. This promise is rooted in the belief that government, when streamlined and focused, can achieve more with less—delivering services that meet the needs of citizens without compromising on quality or accessibility. By embracing efficiency, DOGE has not only reduced waste but also unlocked potential, proving that good governance is not a zero-sum game but a dynamic process of growth and improvement.

Central to this promise is the integration of technology. DOGE's use of artificial intelligence and advanced data analytics has set a new standard for how governments can operate in the digital age. These tools have enabled unprecedented levels of precision in decision-making, from identifying inefficiencies to predicting future needs. They have transformed budgeting from a reactive exercise to a proactive strategy, allowing the government to allocate resources where they are most needed. By automating routine tasks, these innovations have freed up human capital for more strategic endeavors, ensuring that the workforce is engaged in work that truly matters.

Yet, the promise of efficiency is not confined to technological advancements. It also reflects a deeper cultural shift within the government—a shift toward accountability and transparency.

DOGE's participatory initiatives, such as public accountability dashboards and open town hall forums, have shown that efficiency is inherently tied to trust. Citizens, when given a clear view of how their government operates, are more likely to engage with and support its initiatives. This transparency fosters a partnership between the government and its people, turning efficiency into a shared goal rather than a top-down mandate.

The socioeconomic benefits of efficiency are equally compelling. By streamlining operations and cutting unnecessary expenditures, DOGE has redirected billions of dollars toward initiatives that directly impact citizens, such as infrastructure improvements, education, and healthcare. These investments not only enhance the quality of life but also stimulate economic growth, creating a virtuous cycle of prosperity. Efficiency, in this context, is not an abstract ideal but a tangible force for progress, one that uplifts communities and strengthens the nation as a whole.

The promise of efficiency also carries an ethical dimension. By eliminating waste, DOGE has demonstrated a commitment to stewardship—ensuring that public resources are used responsibly and with integrity. This ethical framework is particularly important in an era of growing skepticism toward government institutions. DOGE's efforts show that efficiency is not about cutting corners but about honoring the trust placed in the government by its citizens. It is a reaffirmation of the idea that governance, at its best, is a service rather than a privilege.

Looking forward, the promise of efficiency serves as a guiding principle for the nation's continued evolution. It challenges future leaders to build on DOGE's successes, applying its lessons to new challenges and contexts. It reminds us that the pursuit of efficiency is not a finite project but an ongoing commitment—a promise to future generations that their government will always strive to do better, to be better.

In the end, the promise of efficiency is a testament to the transformative power of thoughtful, deliberate governance. It is a vision of what is possible when innovation, accountability, and the public good come together in pursuit of a common goal. Through DOGE, this promise has been realized, offering a blueprint for a future where governance is not merely functional but exemplary, a source of pride and inspiration for all.

A New Era of Governance

A new era of governance dawned with the transformative work of DOGE, signaling a departure from outdated bureaucratic inefficiencies and ushering in a framework rooted in innovation, transparency, and inclusivity. This era is defined not just by technological advancements but by a cultural shift in how governance is conceived and executed. It is a period where the principles of efficiency and accountability are not mere aspirations but operational realities, shaping a government that is both agile and deeply aligned with the needs of its people.

At the core of this new governance model is the seamless integration of technology. Artificial intelligence and advanced data analytics have redefined the processes of decision-making and resource allocation, allowing for precision that was previously unattainable. These tools have enabled the government to respond swiftly to challenges, anticipate future needs, and ensure that every action is guided by data-driven insights. The result is a government that operates not only more effectively but also more equitably, delivering services that are tailored to the diverse needs of its citizens.

Yet, the new era of governance extends beyond technology. It embodies a fundamental reevaluation of the relationship between government and its people. Through participatory platforms, citizens are no longer passive recipients of services but active contributors to policy development and oversight. Initiatives such as public accountability dashboards and crowdsourced solutions have democratized governance, fostering a sense of shared ownership and responsibility. This participatory approach has not only enhanced trust in government institutions but also strengthened the social fabric, creating a partnership between the state and its citizens.

The era is also characterized by an unwavering commitment to fiscal responsibility. The reforms initiated by DOGE have demonstrated that efficiency and economic prudence are not at odds with one another. By eliminating wasteful expenditures and optimizing resource utilization, the government has not only reduced its debt burden but also reinvested savings into critical areas such as education, healthcare, and infrastructure. These investments have catalyzed economic growth and improved quality of life, ensuring that the benefits of reform are felt across all sectors of society.

Central to this transformation is the leadership model that DOGE pioneered—one that values interdisciplinary collaboration and visionary thinking. Leaders like Elon Musk and Vivek Ramaswamy exemplified how entrepreneurial principles could be harmonized with public service objectives, fostering an environment of innovation and resilience. Their approach has set a precedent for future leaders, emphasizing the importance of adaptability, transparency, and a relentless focus on outcomes.

The new era of governance also embraces the principle of continuous improvement. Recognizing that the challenges of the future will differ from those of today, the government has institutionalized mechanisms for learning and adaptation. Regular evaluations, feedback loops, and pilot programs ensure that policies remain relevant and effective. This iterative approach to governance reflects a humility and openness to change that are essential in a rapidly evolving world.

Perhaps most importantly, this era represents a reaffirmation of democratic values. By aligning efficiency with principles of equity and justice, the government has shown that reform does not have to come at the expense of inclusivity. Instead, it has demonstrated that a government can be both efficient and compassionate, balancing the demands of progress with the needs of its people.

As this new era unfolds, it serves as a testament to what is possible when innovation is guided by a clear moral compass and a commitment to the public good. It challenges other nations to reimagine their own governance models, offering a blueprint for how efficiency can be a force for empowerment rather than exclusion. Through its transformative impact, this era has redefined not only the mechanics of governance but also its purpose, ensuring that government remains a dynamic, responsive, and indispensable pillar of society.

In this vision of governance, efficiency is not merely an operational goal but a guiding philosophy—one that seeks to maximize the potential of government to serve its people effectively and ethically. As the United States continues to navigate the complexities of the 21st century, this new era stands as a beacon of what can be achieved through bold ideas, dedicated leadership, and an unwavering commitment to excellence. It is a reminder that progress is always within reach, provided we are willing to embrace change and pursue it with conviction.

A Call to Action

A call to action resonates across the nation, a summons to engage not only with the progress forged by DOGE but also with the enduring responsibility to sustain and expand its principles. This moment is not just an opportunity for reflection but a mandate to act, to build upon the successes of the past and confront the challenges of the future with resolve and innovation. DOGE's transformative work has laid a robust foundation, but the promise of its vision can only be realized through collective commitment and dynamic effort.

The essence of this call lies in acknowledging the role that every citizen, policymaker, and leader must play in ensuring that the principles of efficiency, transparency, and accountability continue to guide governance. The progress achieved thus far underscores the power of collaboration and the need for ongoing vigilance. Governance is not static; it demands continuous engagement, evaluation, and renewal. DOGE's legacy serves as both a model and a challenge, urging us to think beyond immediate gains and to consider the long-term implications of our actions.

At the heart of this call to action is the imperative to institutionalize reform. The achievements of DOGE demonstrate the potential for government to be both agile and effective, but these gains must be safeguarded against the risks of regression. Ensuring that the principles of DOGE are embedded within the very fabric of governance requires structural and cultural shifts. It demands policies that prioritize accountability at every level, as well as leadership that embraces innovation and inclusivity as cornerstones of progress.

Equally critical is the role of public engagement in sustaining the momentum of reform. Citizens must remain active participants in shaping and overseeing their government. Transparency initiatives, such as public dashboards and participatory budgeting, have shown the power of informed citizenry in driving accountability. These tools must be expanded and refined, fostering a culture where engagement is not just encouraged but expected. The government's success is inextricably linked to its ability to earn and maintain the trust of its people.

Innovation continues to be a vital component of this journey. The technological advancements leveraged by DOGE have redefined the possibilities of governance, but the pace of technological change requires an equally dynamic approach to its integration. Policymakers and technologists must work in tandem to ensure that emerging tools are harnessed ethically and effectively. From artificial intelligence to quantum computing, the next frontier of technology offers opportunities to enhance efficiency while posing new ethical and logistical challenges.

The call to action also extends to addressing inequities and ensuring that the benefits of reform are felt broadly across society. Efficiency must not come at the expense of inclusivity; it must serve as a vehicle for equity. This principle demands a commitment to examining the socio-economic impacts of reform initiatives and making adjustments that prioritize fairness and accessibility. A government that is efficient but fails to serve all its citizens equitably cannot truly be considered effective.

On the global stage, this moment calls for leadership that recognizes the interconnectedness of governance challenges. The success of DOGE has drawn international attention, offering a blueprint for reform that transcends borders. Sharing best practices and fostering partnerships with other nations can amplify the impact of these principles, demonstrating that efficiency and accountability are universal aspirations. This global perspective enriches the domestic conversation, bringing new ideas and approaches to the table while reinforcing the United States' role as a leader in governance innovation.

Ultimately, the call to action is a reminder that the journey of reform is never complete. DOGE has shown what is possible when vision is met with action, but it is the responsibility of every generation to build upon this foundation. This is a call to dream boldly, to act decisively, and to commit unwaveringly to the principles that make governance not just a mechanism of administration but a force for collective good.

In answering this call, we honor the work of those who have come before us and pave the way for those who will follow. Together, we can ensure that the ideals of efficiency, transparency, and accountability continue to shape a government that is responsive, resilient, and reflective of the best of who we are as a nation.

Sustaining the Momentum

Sustaining the momentum of DOGE's achievements requires a continued commitment to the principles that propelled its success: innovation, transparency, and a relentless focus on efficiency. The journey of reform has demonstrated the immense potential of a government willing to adapt and evolve. However, maintaining this trajectory demands vigilance and proactive strategies to ensure that progress does not stagnate or regress in the face of new challenges.

The foundation laid by DOGE provides a robust starting point for ongoing reform. Its integration of advanced technologies, such as artificial intelligence and data analytics, has revolutionized how government operations are conducted. These tools have proven instrumental in detecting inefficiencies, predicting future needs, and allocating resources effectively. Yet, technology alone is not sufficient. Sustaining momentum requires a broader cultural shift within government institutions, fostering a mindset that values continuous improvement and embraces change as a constant.

A critical aspect of this effort lies in institutionalizing the reforms initiated by DOGE. Processes and systems must be designed to withstand changes in leadership and political climates, ensuring that efficiency remains a priority regardless of external pressures. This involves embedding accountability mechanisms into the fabric of governance, from rigorous performance evaluations to public oversight through transparent reporting. By making efficiency a structural imperative, the government can safeguard the progress achieved and build upon it.

Public engagement also plays a vital role in sustaining momentum. The participatory models introduced by DOGE have shown the power of citizen involvement in shaping and monitoring government policies. Expanding these initiatives can deepen trust and ensure that reforms remain aligned with public needs. Citizens must be empowered to hold their government accountable, not only through voting but also through active participation in policy discussions and decision-making processes.

Leadership is another cornerstone of continued progress. The success of DOGE was driven in large part by visionary leaders who brought entrepreneurial thinking and a results-oriented approach to the public sector. Future leaders must embody these qualities, prioritizing innovation and accountability while navigating the complexities of governance. Leadership development programs and cross-sector collaborations can cultivate a new generation of public servants equipped to sustain and expand the principles of efficiency.

Adaptability is equally crucial. The challenges of tomorrow will differ from those of today, necessitating a government that can respond swiftly to emerging issues. This requires a commitment to ongoing learning and innovation, leveraging new technologies and methodologies to address evolving needs. Pilot programs and experimental approaches can serve as testing grounds for future initiatives, allowing the government to refine its strategies in a controlled and iterative manner.

Finally, sustaining momentum demands a long-term vision that balances immediate needs with future aspirations. While the tangible successes of DOGE, such as cost savings and streamlined processes, are significant, they represent only the beginning of what is possible. The true measure of success lies in the government's ability to institutionalize these reforms and create a legacy of efficiency and accountability that endures for generations.

As the nation moves forward, the lessons of DOGE provide both inspiration and a roadmap. They remind us that meaningful reform is not a finite project but an ongoing journey—one that requires dedication, collaboration, and an unwavering belief in the potential of good governance. By sustaining the momentum of these achievements, the government can continue to serve as a force for progress and a source of pride for its citizens, ensuring that the principles of efficiency, transparency, and accountability remain at the heart of the nation's democratic ideals.

www.ingramcontent.com/pod-product-compliance
Lightning Source LLC
Chambersburg PA
CBHW061725270326
41928CB00011B/2111